Feedback Loops

Postphenomenology and the Philosophy of Technology

Series Editors: Robert Rosenberger, Peter-Paul Verbeek, Don Ihde

As technologies continue to advance, they correspondingly continue to make fundamental changes to our lives. Technological changes have effects on everything from our understandings of ethics, politics, and communication, to gender, science, and selfhood. Philosophical reflection on technology can help draw out and analyze the nature of these changes, and help us to understand both the broad patterns of technological effects and the concrete details. The purpose of this series is to provide a publication outlet for the field of philosophy of technology in general, and the school of thought called "postphenomenology" in particular. The field of philosophy of technology applies insights from the history of philosophy to current issues in technology and reflects on how technological developments change our understanding of philosophical issues. Postphenomenology is the name of an emerging research perspective used by a growing international and interdisciplinary group of scholars. This perspective utilizes insights from the philosophical tradition of phenomenology to analyze human relationships with technologies, and it also integrates philosophical commitments of the American pragmatist tradition of thought.

Recent Titles in This Series

Feedback Loops: Pragmatism about Science & Technology, edited by Andrew Wells Garnar and Ashley Shew

Sustainability in the Anthropocene: Philosophical Essays on Renewable Technologies, edited by Róisín Lally

Unframing Martin Heidegger's Understanding of Technology: On the Essential Connection between Technology, Art, and History by Søren Riis, translated by Rebecca Walsh

Postphenomenological Methodologies: New Ways in Mediating Techno-Human Relationships, edited by Jesper Aagaard, Jan Kyrre Berg Friis, Jessica Sorenson, Oliver Tafdrup, and Cathrine Hasse

Animal Constructions and Technological Knowledge, by Ashley Shew

Using Knowledge: On the Rationality of Science, Technology, and Medicine, by Ingemar Nordin

Postphenomenology and Media: Essays on Human–Media–World Relations, edited by Yoni Van Den Eede, Stacey O'Neal Irwin, and Galit Wellner

Diphtheria Serum as a Technological Object: A Philosophical Analysis of Serotherapy in France 1894 - 1900, by Jonathan Simon

Feedback Loops

Pragmatism about Science and Technology

Edited by
Andrew Wells Garnar and Ashley Shew

Afterword by
Joseph C. Pitt

LEXINGTON BOOKS
Lanham • Boulder • New York • London

Published by Lexington Books
An imprint of The Rowman & Littlefield Publishing Group, Inc.
4501 Forbes Boulevard, Suite 200, Lanham, Maryland 20706
www.rowman.com

6 Tinworth Street, London SE11 5AL, United Kingdom

British Library Cataloguing in Publication Information Available

Library of Congress Cataloging-in-Publication Data

Names: Garnar, Andrew Wells, 1974- editor. | Shew, Ashley, 1983- editor.
Title: Feedback loops : pragmatism about science and technology / edited by Andrew Wells Garnar and Ashley Shew ; afterword by Joseph C. Pitt.
Description: Lanham : Lexington Books, [2020] | Series: Postphenomenology and the philosophy of technology | Includes bibliographical references and index. | Summary: "This volume explores the arrangement of science, technology, society, and education. Using the concept of 'feedback loop', this book processes subjects dear to the work of Joseph C. Pitt: technology as humanity at work, pragmatism, Sicilian realism, pragmatist pedagogy, instrumentation in science, and more"— Provided by publisher.
Identifiers: LCCN 2020034182 (print) | LCCN 2020034183 (ebook) | ISBN 9781498597623 (cloth ; permanent paper) | ISBN 9781498597630 (epub)
Subjects: LCSH: Technology—Philosophy. | Pitt, Joseph C.
Classification: LCC T14 .F39 2020 (print) | LCC T14 (ebook) | DDC 601—dc23
LC record available at https://lccn.loc.gov/2020034182
LC ebook record available at https://lccn.loc.gov/2020034183

Contents

Preface

As we have worked together to co-edit this volume, we've been grateful for a community of scholars and friends who have enthusiastically agreed to contribute, to offer suggestions, to ask when the book party will be, and to engage with and revisit important work in our field. We offer this volume not as a *festschrift* but as a work in Philosophy of Technology, Philosophy of Science, and Science and Technology Studies (STS) that lends a sometimes-critical eye and a sometimes-appreciative glance at the breadth and meaning of the work of Joseph C. Pitt. Joe, as most of us know him, has been a provocateur of the best sort—asking for (demanding, loudly and at least once while standing on top of a chair) straightforward accounts, developing critical ideas that force others to sharpen their wits, and, through his pugnacious and Socratic questions, making us all better for the clash, even when we remain in disagreement.

Do you know that scene from the comedic movie *Dodgeball* where the old man who is going to train the hapless, underdog dodgeball team of misfits yells, "If you can dodge a wrench, you can dodge a ball!" then whips out a wrench and throws it at the group? Being a colleague or student-colleague (for that's how he treats us) of Joe's can feel like this: he's always ready with a counterexample or some other wrench to throw at you. He does this in a good-natured way—sometimes the counterexamples are fun and funny—and you may laugh along the way to learning to be more agile in thought, but it's a wrench nonetheless. We feel like he's done this throughout his career and for the whole of philosophy of technology and STS as disciplines. And we're better for it—and sometimes we're both grateful and crabby about it, too.

We thank everyone in our community for supporting this volume—whether through the actual writing of chapters or just enthusiasm. It's overdue, and we've gotten that feeling from everyone we've encountered. We thank our authors for doing the hardest of work—dodging those wrenches and still wanting to write about them. We take the name of this volume from an idea central in Joe's work: feedback loops. Feedback loops, as used in Joe's work, point to the iterative nature of engineering design: loops of successive feedback help shape a technology or system. We offer this volume in the same vein: here we have a few loops of feedback on work in STS and philosophy of technology influenced by Joe's work.

LET'S SET THE CONTEXT: PHILOSOPHY AS A HYPERSPACE TRANSMOGRIFIER?

Contemporary technoscience can seem baffling. Almost like science fiction. This might help account for why so much of Science and Technology Studies reflects an out-of-this-world ethos, invoking terminology that echoes this situation. There is talk of "hybridity," "cyborgs," "extended cognition," "the anthropocene," and more. Such terminology reflects the tenor of the times in which we live: where big science is technologized, and all the problems are thorny.

Hybridity and networks are two of the dominant motifs of science and technology studies, and with good reason. Technoscience is about making *strange connections*, whether these be the fusing of unlike objects (GMOs), linking orders of magnitude (from neuroscience to over behavior), opening new modes of perception (the Hubble Telescope), etc. One way to make sense of contemporary technoscience is in true science fiction terms. What we have in mind here is what is necessary in a philosophy that operates as a science fiction device, like the kind that drives a text (like William Gibson's *Peripheral* in the eponymously titled novel). In a spirit that echoes both Bugs Bunny (in particular the episodes where he dealt with Marvin the Martian) and Calvin and Hobbes, technoscience operates as a "hyperspace transmogrifier." It transforms while also shrinking distances. Unlike either Calvin's transmogrifier (or, if memory serves, Marvin's devices), it is not a box that one places over the object to be transmogrified. Instead, it functions in-between objects, making unexpected linkages. After all, it is a *hyperspace* transmogrifier. Where is the fun in hyperspace if the objects connected have to be right next to each other?[1]

Although Joseph C. Pitt would be loath to being described in such a way, where his writings have ended up reflects a similar postmodern, science-fiction frame of mind. Of course, he scrupulously seeks to avoid the excesses of postmodern science and technology studies, and much of his writings beginning around 1980 tend to eschew the excesses of analytic philosophy. A "Sicilian realist," Pitt cuts away bullshit and wants pragmatic explanation; he brings this to bear in how he explains the organization of science and technology, the way in which theories change through our use of technologies, about scientific explanation versus technological explanation, about the nature of technological knowledge, about the politics and politicization of science and technology, and how we define what technology is so we can develop better accounts of technology that don't categorize it by near-mystical essence, aura, or paradigm. Rather, Pitt pushes his readers, his students, and his colleagues to take technologies on individual terms. We're not allowed to talk about science in the singular—there are sciences—and we're not allowed to talk about technology in the singular either—there are technologies. When we get to these overarching concepts and talk about them in the abstract, it's a tendency of philosophers, analytic and continental alike, to make sweeping and general claims that don't actually map onto the use of individual technologies. As an educator, he's full of examples that push students to reconsider their sweeping claims and to develop better accounts. Pitt pushes for a Heraclitian view of technology (and everything else). Echoing Heraclitus, the pre-Socratic philosopher who held that everything was change, Pitt would have us think that there is no one overarching narrative to tell about technologies in our lives and world—and stands in firm opposition to any theorist who would be so sloppy; he's known for his pugnaciousness.

While arguably old hat in some philosophical quarters (thinking here of anyone who traces their ancestry back to Nietzsche, whom Pitt does have a soft spot for), Pitt's Heraclitianism is revolutionary within Anglophone philosophy of science. Yes, Thomas Kuhn laid the groundwork for this movement. Yet, Pitt avoids what so many philosophers of science sought to in the wake of *The Structure of Scientific Revolutions*: to find permanent structures beneath the churning chaos of the history of science. His earlier works, like *Pictures, Images and Conceptual Change: An Analysis of Wilfrid Sellars' Philosophy of Science* (1981) and *Galileo, Human Knowledge, and the Book of Nature: Method Replaces Metaphysics* (1992), still held to the perennial hopes, but this was all but abandoned in his 2000 *Thinking about Technology*.

In the same vein of change and heterogeneity, Pitt refuses to use the very popular STS term "technoscience" in his own work, preferring to keep separate "science" and "technology." He avoids simply reductionist answers, usually choosing to embrace complexity (which arguably he learned from his colleague Marjorie Grene and grew out of discussions with Richard Burian). Instead of expecting to find stability across time, Pitt has come to embrace a Heraclitian philosophy that emphasizes change, be it material, conceptual, social, or any other.

Among his most common moves is to embrace *deflationary definitions*, but with less he gains more. Probably the most infamous of these is his definition of technology: "humanity at work" (Pitt 2000). This short definition is at once provocative and productive: we are now forced to think about humanity's work, in all its forms. While Pitt does end up shifting much of what humans produce into the category of "technology," it would be a mistake to think of this as a "reduction." He is not seeking to explain the operation of all technology based upon supposed essence. "Humanity at work" does not function in that way.

For example, Andrew remembers a presentation Pitt made to the Virginia Tech Science and Technologies Studies program. At the end of it, Richard Burian challenged Pitt over his definition of technology. Burian said, almost shouting, something to the effect of "this is the same thing I've asked you for years: what doesn't count as technology?" Burian's frustration is a typical response to the vagueness of Pitt's definition. At first pass, it does very little to clarify what counts as technology and how technology works.[2] Pitt's response to Burian was telling: in roughly the same tone, he said, "That's the point." Beyond some rather generic claims dealing with feedback loops (which are actually largely normative), Pitt does not think that much can be meaningfully said about "technology" *simpliciter*. Outside of the vacuous observation that technology is the product of goal-directed human activity, Pitt does not think that a great deal more can be said that applies to *all technology*. He seeks to resist here what he associates with Heidegger and Ellul: grandiose claims about how all technology operates (or at least all *modern* technology). Even when Pitt's interpretations of essentialist philosophers of technology skew toward the simplistic, Pitt has a point. Finding a common, *informative*, thread that runs among all technologies seems like a lost cause. This seems especially true if one looks over the full sweep of human history, but even if one looks at modern technology, it appears plausible.

The point of such a broad definition is to break bad habits of trying to establish reductive essences, and instead turn philosophers toward what Pitt finds actually fruitful: dealing with the specifics of particular technologies. This is representative of what is called "the empirical turn" in the philosophy of technology. In its early years, many philosophers who dealt with technology tended to write similar to Heidegger: in very broad strokes. This work served the purpose of at least establishing that there are important, distinctively philosophical questions, even if it tended too general to provide much concrete insight.[3] Against this trend, philosophers like Pitt, Davis Baird, Alfred Nordmann, Rachel Laudan, Diane Michelfelder, Allan Franklin, and Ian Hacking, along with many historians of science and technology like Ann Johnson and Peter Galison, turned to the details of how actual technologies actually functioned—in both a sense of individual device function and their function within a system or infrastructure. These philosophical discussions helped to shape his definition, and the understanding he presents of the sciences, technologies, and humanity along with Pitt's debt to pragmatism.

As the chapters in this collection demonstrate, Pitt's thought is not necessarily about settling arguments in a "once and for all" fashion. Officially, he would prefer that the interlocutor be convinced by the arguments in favor of his position. In reality, Pitt would likely find such agreement boring. One value of his work is not for what it closes down, but opens up. Engaging with Pitt, whether as a student, a colleague, or a reader, moves one into new spaces. Again, a function of the hyperspace transmogrifer. Regardless of where one is at home, Pitt moves you to somewhere else. Yet, when he engages with you, he is changed as well. A transmogrifer with a feedback loop.

Another memory. Andrew was talking with Pitt after Joe had agreed to chair another dissertation committee. Pitt's comment about this: "I'm going to learn a lot." Unlike many advisors who seek to imprint themselves on their students, Pitt genuinely worked with them in a relationship of open exchange: a feedback loop. He required the student to perform at the highest level (how we both got through is still a mystery), and Pitt did have a worn rubber stamp emblazoned with "bullshit" on it. This helps explain why so many of his students wanted to contribute to this volume.

On a personal level, it is clear that Pitt relished the joust, the clash of ideas. Despite his criticisms of Rorty (not undeserved) of an endless (philosophical) conversation, this is something Pitt obviously enjoys. There are rules, first and foremost: be clear (part of his criticism of Heidegger and Derrida: why this does not apply to Sellars is not obvious). From here, indefinitely long discussions about philosophy can ensue. Even if Pitt does not acknowledge it, his positions will change through such engagements. These discussions are much closer to Peirce than Rorty, in that, if carried on long enough, they will have some sort of the stable result. Furthermore, much like Peirce, it is only through conversation that this can be achieved (odd for Peirce, a near hermit in his last years, but not for Pitt who seems to know so many academics across the globe).

We've arranged these chapters, from both colleagues and students of Pitt's, in order to map first the challenges of science and technology, and then beyond. In that vein, we have the following chapters:

- Anne Fitzpatrick, a former student and current government employee, writes on high performance community and national competitiveness;
- Ronald Laymon, emeritus philosophy professor and legal scholar, writes on copyright issues in synthetic biology;
- Andrew Wells Garnar, former student and one of the editors here, provides a defense of Sicilian realism, a brand of philosophy developed by Pitt;
- Nicholas Rescher, notable philosopher of science—and also once a professor of Pitt!—explores fictional idealization as useful in communication;

- Ashley Shew, former student and another of your editors here, writes about disability technologies through the lens of technological knowledge;
- Ivan Guajardo, another former student and current philosophy professor, examines social networking sites to think about the neutrality or non-neutrality of technology;
- Tom Staley, engineering professor and former student, reflects on pragmatism and the approach of Joe Pitt;
- Peter Kroes, philosopher of technology and long-standing Pitt interlocutor, revisits moral values as embedded in technologies (contra Pitt, who holds technologies as neutral);
- Brandiff Caron, former student and current engineering education educator, reflects on pragmatism within engineering education;
- Allan Franklin, emeritus historian and philosopher of science, provides two important historical case studies of experimental results made possible through technology; and
- Jim Collier, student, friend, and colleague who works on social epistemology, provides our last chapter and reflects on Pitt's approach and significance as an educator.

Finally, this volume concludes with a short reflection from the Sicilian Realist himself, Joseph C. Pitt (who variously gets called Joe and Pitt through this volume). He reflects mainly on his teaching, but his teaching and his philosophical work are deeply entwined: he shows us his methods and ideas in his approach to science, engineering, and human projects. Pitt has spawned thinking on such a wide variety of topics, as seen in the list of author contributions above. His students work for the government and for universities, as well as in music and in small business (not as well represented here, but true nonetheless). His colleagues hail from multiple departments and come with disparate interests. Even when Pitt is bewilderingly wrong on matters (see Kroes' contribution), his hyperspace transmogrifying feedback loop connects, transforms, and relocates what it engages with in important ways.

All of this is to say that by following the various paths that Pitt's thought opens up provides a unique and valuable way to understand contemporary technoscience. The authors gathered here trace out these strands, sometimes in agreement with Pitt (in particular or in spirit) and sometimes opposed, illuminating distinctive features of science, technology, society, and the relationships between them. Among other things, one value of Pitt's work, especially once he began publishing on technology in 1980, is that he takes seriously the way society can shape technology and science. When compared with more traditional philosophers of science, this should be seen as relatively radical. Yet, even if he is not radical enough in drawing the science/society feedback loop, Pitt does sidestep the standard criticisms that can be leveled at STS scholars and postmodernism. For this reason, we consider his scholarship an important link between communities and topics, as well as central to philosophy of technology as a discipline, and as a discipline concerned with instruments, devices, the sciences, and politics.

<u>Andrew Wells Garnar and Ashley Shew</u>

NOTES

1. Andrew admits that he wrote this while listening to "The Carl Stalling Project," which is a selection of Stalling's Warner Bros. cartoon soundtracks, written between 1936 and 1958.

2. As Shew has argued (2017), those things that are relatively clear are open to criticism, especially the limit on technology as a humans-only phenomenon. Another clue as to what does not fall under his definition has to do with the concept of "play."

3. In actuality, this more general approach did a great deal of important work. There is also a question about whether such work is still necessary. Pitt, who tends to be more nominalist than he should, usually rejects it out of hand.

Chapter 1

The Pursuit of Machoflops

The Rise and Fall of High-Performance Computing

Anne C. Fitzpatrick

Nearly twenty-eight years ago, I first encountered the writings of Carl Hempel, Rudolf Carnap, Sir Karl Popper, and other members of the Vienna Circle school of philosophy in Joseph C. Pitt's mandatory-for-science and technology studies (STS) philosophy graduate course at Virginia Tech. Reading Popper's *The Logic of Scientific Discovery* (1968) was hard going the first time.[1] The logical positivist syllabus that comprised Joe's class was the most daunting one in graduate school, with a paper due weekly; students largely analyzed the Vienna Circle school of philosophical thought and some other, more modern writings about science and technology. Carl Hempel's tome, *Aspects of Scientific Explanation* (1965), was a much-dreaded-by-students rite of passage to get through Joe's first-year course. But it challenged you and made you stronger intellectually and pushed your writing skills. If you could get through the logical empiricist philosophers, you could get through any course. Being more a student of history than philosophy, I subsequently read Popper's disquisition, *The Open Society and Its Enemies* (1994), on my own to try and understand what the Vienna Circle really stood for.

Today, nearly a century after they were written, the works of Popper and several of his other Viennese intellectual contemporaries are again becoming increasingly relevant. There is a renewed appreciation for their writings now in a time when our own democracy seems at risk, and science—especially climate change—is dismissed as political and unfounded, and the United States government is diminishing its general support for science. What I did not understand when initially exposed to the works of Popper, Quine, Carnap, and Hempel was that their writings were essentially a response against the rising tide of fascism and tyranny in Europe in the 1930s.

Many members of the Vienna Circle had to flee their beloved Vienna and emigrate abroad to find safety. Popper wrote *The Open Society* after Hitler invaded Austria. In *The Open Society*, Popper advocated that central direction is not the way to govern a society, but rather competition for ideas supported by critical thinking would lead to a better way of life. A lot of his beliefs were based on his earlier works that explored the scientific method where hypotheses are produced and advanced, and scientists try to falsify them, so any hypothesis that remains unfalsified must stand as some kind of credible knowledge. This notion carried forward into a concept of truth for Popper in *The Open Society*, where he argued that blind deference to great men and grand theories of history was dangerous to modern, civil, active participant society.

In this chapter, I will review and analyze a slice of the high tech government-industry partnership that builds America's high-performance computers (HPC, also called supercomputing), where I have worked for some time, and demonstrate how Joe's work in the philosophy of technology has deeply influenced my approach to my job. As part of a modern, civil, and open society and as a federal employee and leader, I need credible knowledge to make the most informed decisions I can when spending public money. Because of this, I find myself thinking a lot not only about Joe's work, but also about Popper and the logical empiricists' writings. I do this both in terms of their relevance to our American democracy today, and also in seeking credible scientific and technological knowledge in my daily work over the past two decades in the information technology and national security and policy realms, which when executed correctly is intended to support our democracy.

The logical empiricists' writings were not without their flaws, as postmodern and social constructivist scholars in the middle and latter twentieth century pointed out. In the STS coursework, we learned that women and minorities were usually left out of histories, and perhaps science and technology could be seen through relativistic lenses and were up to interpretation (Haraway, 1991; Rossiter, 1982). STS, as an outgrowth in part of Philosophy of Science, has come a long way as a discipline since academics began to question what science is, how it is conducted, and where science and technology fit into society and what their relationship to one another is. Now, more than ever, we need to figure these relationships out.

My STS education, the logical empiricists, and other philosophical points of view influenced my choice of career, while Joe's philosophical writings about technology, in particular, have shaped both my analytical work on technological futures and my efforts to bring closer together the "two cultures" of—for our purposes here—Science, Technology, Engineering, and Math (STEM) and non-STEM specialists (Snow, 1962). For some years I've supported efforts to help the United States government avoid technological surprise to ensure that we remain on the leading edge of HPC. My office colleagues and I have watched how over the past couple of decades China has been making huge strides in HPC, even surpassing the United States at times in sporting the fastest machines on the planet. Until November of 2018, China had the fastest supercomputer in the world: the TaihuLight, at the National Supercomputing Center in Wuxi. TaihuLight's microchips are all Chinese-designed—a landmark achievement for China, considering that prior to 2000 it had no computers on the biannual and increasingly questionable Top500 ranking of the fastest HPCs in the world.[2]

Today, China has 226 HPCs on this list. I'll return to the subject of HPC as a real-life STS case study later in this chapter, but first, I need to provide some context based on Joe's ideas about technology and the human condition. Advances in supercomputing matter a great deal both as they are an indicator of a nation's economic competitiveness and because of how pervasive information technology has become. Because of this, I find myself thinking about Joe's notion that as a society we are all dependent on a technological backbone, which is rapidly becoming more and more of a digital backbone that weaves through nearly all facets of modern life (Pitt, 2000). Joe would even say technology—or just "Tech" as it is increasingly called—is more than pervasive in modern life, where it defines the human condition today (Pitt, 2000).

Joe was right. It's easy to see the pervasiveness of tech where the speed of change and innovation has become so blindingly fast. Joe foresaw what is becoming obvious to all now—that tech is a way of life today, especially in more affluent communities and parts of the world. From smart fridges in our homes to driverless cars in cities and our phones essentially extensions of our bodies and personalities, technology defines the human condition in the developed world: we're often expected to have work-issued phones and laptops, and to be available during nonstandard hours; we come home at the end of the day and without a thought check work email and order pizza via Amazon Echo without touching a button or talking with a real person. For many people, there is little "turning off" from all this unless one deliberately and mindfully chooses to, and in today's world that is not always easy given the dependence we have on these systems.

The last twenty years or so have witnessed a critical transition from the twentieth-century US-led military-industrial complex, which is by no means dead, to the globalized Big Tech economy, dominated by Silicon Valley and its Chinese competitors and suppliers. Facebook, Amazon, Apple, Netflix, and Google (acronymized as the FAANG stocks by investors) are collectively valued at over $3 trillion as of 2018. At the time of writing this chapter, these and other competitors are sitting on unprecedented amounts of cash. These companies' influence on our daily lives is unmistakable, both providing convenience and creating serious risks.

Big Tech is no stranger to all kinds of ethical issues, such as issues of privacy and surveillance including hackers spying inside homes via unsecured baby monitors, and theme parks requiring every guest to register themselves and wear a tracking bracelet during their visit so their every stop, purchase, and move can be recorded and processed in a cloud-based computer system, creating a data-driven personal profile (Marsden, 2019). Such phenomena demonstrate that Tech is not neutral; if humans made it, that tech has a human imprint on it in some way. For example, artificial intelligence (AI) programmers are discovering that their machines can carry the racial and gender biases of those who created them, even though the programmers did not likely intentionally program their AI in that way. As Joe has pointed out repeatedly throughout his career, technology is not inherently evil or destructive, but technological change is a complex set of events that can have far-reaching economic, physical, and psychological effects on people, and it's expected that there is a fear of the unknown consequences that will arise from new and evolving Tech (Pitt, 2000).

At this point in time, no one truly understands how deep neural networks (DNN) as part of an AI system "learn" to recognize images, words, and other data points. It is important to point out that AI itself is not intelligence; it is a collection of computational techniques used to help us close the performance gap that is stressed by high-speed computing. AI techniques always trade search algorithms for knowledge. Using words like "learning" and "deep" doesn't make it so. Even if algorithms are tuned to a great data set for better precision and speed, they may be confused by a changing environment, averaging multimodal data, sampling errors, and assumptions about what is happening. And presently, in a commercial sense, AI is a buzzword widely used by vendors to sell often useless products to the government.

Joe's philosophical view of AI is more grounded. He has characterized the goal of AI as figuring out the mysteries of human cognition and allowing for new paths leading from human sensory input to coherent thinking and knowledge, resulting in radically new outputs (Pitt, 2017). AI right now is a fascinating journey, not an endpoint.

In most cases, technology is inevitably entangled in politics and often driven by personal and corporate interests. A classic example—one that Joe always had a keen interest in and wrote a great deal about—is the famous Italian astronomer Galileo Galilei, who is probably best known in today's popular culture as the revolutionary scientist who invented the telescope and was branded a heretic by the Catholic Church—essentially a powerful corporation—after proclaiming that the Copernican theory of the universe, where the planets revolve around the sun, was correct. Most people—and authorities—at that time believed in an earth-centric universe. Galileo eventually wound up dying under house arrest in 1642 for teaching and defending the

Copernican system. Nevertheless, Galileo's innovative telescope lived on, was improved upon, played an important role in changing the way humans perceived their place in the universe, and had long-reaching impacts on society and science (Pitt, 1992).

Out of all his writings, I've found Joe's *Thinking about Technology: Foundations of the Philosophy of Technology* (2000) the most useful when helping me to think through and solve problems at the office because it describes many practical applications toward real-life technology quandaries. In *Thinking about Technology*, Joe defines technology broadly as humanity at work, with work being "the deliberate design and manufacture of the means to manipulate the environment to meet humanity's changing needs and goals" (Pitt, 1992). Importantly, Joe establishes that technology is a form of knowledge that does not necessarily have to be tied to or dependent on science and should not be considered as inferior to the latter.

Tech's not taking a second place to science is nowhere more evident than in the twentieth and twenty-first centuries. During this time, I would argue that there has been no technology—not spacecraft, jet engines, or nuclear weapons—that has had as large an overall impact on humanity as the computer. *Computing is the technological backbone undergirding modern life today.* As part of the Big Tech economy, computers are an inextricable part of our society, homes, relationships, and culture; they underpin commerce, medicine, finance, education, research, and most everything else we engage in. How we even define what a computer is remains up for grabs: their capabilities have been advancing incrementally for decades and they have taken on many forms, from game consoles and handheld smartphones, to larger than football field-sized data centers. Computing is arguably the backbone of our global economy today, and semiconductor products are the third-largest class of U.S. exports. Recent research has shown that one-third of productivity increases in the United States since the early 1970s came from the broad field of computing (Byrne, Oliner, and Sichel, 2013).

Modern computing has its roots in World War II and has evolved extensively since then, growing steadily but slowly in terms of memory and performance. Very large, general-purpose, and powerful supercomputers were once almost exclusively the realm of secret government defense laboratories. Specifically, after World War II, the Los Alamos and Lawrence Livermore National Laboratories as part of the U.S. Department of Energy (previously known as the U.S. Atomic Energy Commission) became key sponsors of and customers for specialized scientific HPC systems. They quickly established the speed of a machine's floating-point arithmetic operations as the performance criterion defining supercomputing (MacKenzie, 1991). Incredibly, over seventy years later, this is still primarily how HPC performance is gauged: specifically, speeds are measured in Floating Point Operations Per Second (FLOPS).

While the first "supercomputers" such as the Electronic Numeric Integrator and Calculator (ENIAC) boasted speeds of around 500 FLOPS, today we have reached the petascale or petaflop (PF) era, with machines that can run one quadrillion FLOPS.

HPCs are still found in government settings, but today they are also in major research universities and in certain commercial industry laboratories. In these settings, they solve all sorts of specialized scientific and engineering problems ranging from oil exploration to climate change, and on Wall Street automatically execute millions of daily stock market trades often without many humans in the loop. Yet, HPCs have largely remained unadopted by middle- and lower-end users, which is unfortunate because of HPC's potential benefits to them; I will return to this subject later and explain it more thoroughly.

As with Galileo's telescope, scientific instruments can make fascinating case studies of human faith in and their relationships with technology. HPC is an interesting case study in technological advancement because it's a very specialized, even rarified, segment of the Big Tech economy; yet, HPC is embroiled with political wrangling, big egos, personal ambitions, and curious behavior where many of its practitioners compete to be able to boast of owning the fastest HPC on the planet. Japan currently holds this title: as of June 2020, the world's fastest HPC, Fugaku, located in Kobe, demonstrated a speed of 415.5 PF. At that time, the US held the number 2 position with Summit at 148 PF, while China's TaihuLight placed number 4 on the Top500 list at 93 PF Beyond this, HPC builders are now jockeying to be first to reach exascale computing, machines capable of at least one exaFLOPS, or a quintillion (or 10^{18}) calculations per second. Yet, HPC builders are incorporating very little fundamentally new technology into the design of these devices, instead scaling out from what technology they have already and with the limited choice of products that the supercomputer vendors will provide.

While the Top500 list is neither the only nor the single most important measure of HPC's advancement, these numbers do signal two important trends: first, that the United States is no longer significantly far ahead of foreign competition and, second, that HPC has become an inextricable and commoditized part of a global industrial ecosystem that is rapidly changing. China's ascent in HPC is an important bellwether for significant shifts in the global economy and international security.[3]

Times have changed but HPC practitioners have not. The Top500 project was started when it became clear in the early 1990s that a new definition of supercomputer was needed to produce meaningful statistics, intending to provide a reliable basis for tracking and detecting trends in high-performance computing and bases rankings on a portable implementation of the Fortran-based LINPACK benchmark for distributed-memory computers.[4] Every machine that is measured for the Top500 list is given the same LINPACK benchmark to run.[5] Lately, detractors have vocally criticized the LINPACK benchmark, saying that it has succeeded only because of its scalability, and for showing unrealistic performance levels that would generally be unobtainable by all but a very few programmers who optimize their code for their specific machine, because it only tests the resolution of dense linear systems, which are not representative of all the complex operations usually performed in scientific computing. Most HPC builders today recognize the faults with the Top500 list, but continue to adhere to it; changing ingrained culture and established ways is always challenging.

The reality is that there has been too much emphasis among supercomputer builders and their government sponsors on merely attaining the highest score on the list, with little regard for the actual quality or true performance of the machine when it tackles real-world problems. Within the HPC community this superfluous chase to be number one on the Top500 list is today known unofficially as the quest for "machoflops," where HPC builders and their sponsors boast in nerd-speak that "my machine has more FLOPS than yours," or "my HPC is bigger than theirs." According to HPC practitioners, the term "machoflops" has been in colloquial use since the 1980s, where allegedly vendors such as the (now defunct) Thinking Machines Corporation, IBM, and others—in order to undercut each other—would accuse competitors of overhyping that their products were the fastest available for purchase. Machoflops never went away, and neither did the Top500 list.[6] Alternatives to the LINPACK benchmark do exist, such as the Green500 list, which measures HPCs in terms of energy efficiency. Such alternative rankings, however, are not that widely accepted by the HPC community. Why not? The answer is a complex mix of human decisions and values, technology, politics, and economics.

HPC technology is based on long-established, complementary metal-oxide semiconductors (CMOS), sometimes referred to as "classical computing," as opposed to over-the-horizon concepts like quantum or superconducting computing, which are still in the basic research stages and may not be realized for some time. A looming problem for CMOS is the likely death of semiconductors as we know them: since coined by Intel Corporation founder Gordon Moore in 1965, Moore's Law states that the number of transistors in a dense integrated circuit doubles about every two years. Moore's Law has held true for a long time, but if experts in the semiconductor industry are correct and we continue to cram ever more transistors on each chip, we're going to run out of room on the silicon slices. To be clear, Moore's Law is not a law in any legal sense and is more akin to an observation and projection of technological change. It is viewed by some as a human construct. Philosopher Cyrus Mody has even suggested that Moore's be viewed as a *regula*, which

> conveys multiple senses that reflect the heterogeneity of Moore's Law and phenomena like it: it is a rule to be obeyed; it is something made with a ruler, i.e., a human construct that straightens out complexity; it is a regularity observed in the world; and it has a regulatory function. (Mody, 2017, 242)

However we characterize Moore's, along with the unraveling of it and closely related, Dennard Scaling is also breaking down. Named after a 1974 paper coauthored by Robert Dennard, this "law" states, roughly, that as transistors decrease in size, their power density stays constant, so that the power use stays in proportion with area (Dennard et al., 1974). But since around 2005 the reduction in the size of transistors associated with a decrease in power has ended, leading to an inability to increase clock frequencies up as the transistor size decreases. The inability to operate within the same power envelope led the HPC industry to transition to multicore architectures, creating significant challenges for memory technology where with each new generation of Central Processing Unit (CPU) the number of memory controllers per core decreased and the memory system burden increased.

In the meantime, HPC component technology became a broad set of globally produced and desired commodities, while worldwide competition has stiffened to outcompute other nations, and almost anyone with the resources can build an HPC, which means most developed nations. U.S. government and private investments in HPC have been inconsistent in the past decade, whereas investment in data analytic-oriented or data computing—driven by consumer demand for AI-enabled smartphones, gaming, and scores of other personal consumer products—has skyrocketed. There is far more money to be made in data computing; therefore, the industry is chasing that more than government HPC projects. The supercomputer industry and the government both failed to foresee the rise of data-oriented computing and its seemingly limitless ever hungry for cool, new devices customer base.

For well over two decades HPC architectures have strayed further and further away from an optimal balance between processor speed, memory access, and input/output (I/O) speed. Since 1991, successive generations of HPC systems have upped peak processor performance without corresponding advances in per-core memory capacity and speed, and thus the systems have become increasingly compute-centric and the well-known "memory wall" in these devices has gotten worse. This means that every time the HPC industry comes out with a new flagship machine it is in practice slower than the previous generation: what this means is that when a real-world complicated code representing a hard science problem is run on the machine, users get single-digit efficiencies and much slower speeds than when tested by LINPACK. HPC designers have made a few noteworthy technological advances in the past two decades, such as message passing interface (MPI), but not significantly much else.

Many complex scientific problems are best solved by being parallelized; HPC parallel computing is accomplished by splitting up large and complex tasks across multiple processors. The 1990s saw the introduction of Linux and other open-source software that could run on inexpensive, off-the-shelf PCs, and this work opened the door for low-cost, high-performance cluster computing. New standards and tools were also developed that allowed for distributed memory parallel computing systems that made it easier for programmers and a variety of scalable and portable parallel applications.

The advance of parallelism has been a boon to scientific problem sets, such as modeling nuclear weapons and understanding supernovas. But twenty and some years ago, HPC practitioners' fundamental assumptions about applications and system software did not anticipate exponential growth in parallelism. The number of system components increased faster than the component reliability, while undetected error rates continued to increase alongside failures. Increasing power requirements are driving HPC builders to use more processors instead of developing newer and faster ones. This is a vicious cycle: while US vendors are struggling to reduce chip power, their solutions often increase fault rates and involve compromises in capabilities.

Besides the technical obstacles and lack of foresight, there are social and political reasons for the near halt in HPC's evolution: the government has steadfastly refused to invest in much new and alternative-to-CMOS computing technologies, instead insisting on mere scale-out of what vendors already provide. HPC industry vendors such as Intel, Cray, and IBM will not make the investment in either new technology or software as they are too focused on quarterly profit results instead of long-term research investments. And, as we saw earlier, HPC specialists and some government sponsors would often remain too focused on vanity projects that would propel them to the top of the Top500 ranking, rather than build the most efficiently usable supercomputers.

Witnessing all this unfold over a number of years, I find myself looking back to how Joe taught me that thinking philosophically can help us make sense of complex problems such as the state HPC has found itself in. Putting my STS hat on, I ask the question: why do we even want or perceive we want "the best" when it comes to technology? Is it the American, workaholic, competitive ethos? Perhaps somewhat, but it is all these factors I've just described, coupled with the reality that the United States does not have the political support to do this because HPC remains invisible to most policymakers who are in positions to have an impact on it.

The nature and culture of the HPC research and development community also factor into where this industry stands today. This is not to assign blame on individuals, but taking an applied STS view of how this failed, some description of the HPC designer culture and community is necessary to fully understand the big picture. HPC is a nerdy, niche, overwhelmingly over-age-fifty male segment of the tech landscape—the realm of an elite set of weapons and rocket scientists who live and work in a vacuum and have no reason to consider how HPC could benefit the larger world.[7] When attending an HPC conference, one notices that the field does not attract many young (meaning under age forty) professionals, who are today drawn to the excitement of data computing based on newer languages such as Python, SQL, and others which are easier for new developers to learn and master than programming HPCs.

In *The Soul of a New Machine* (1981), author Tracy Kidder chronicled the team of engineers in the then startup Data General Corporation. The team worked under enormous pressure and at a furious pace to complete the Eclipse MV/8000 computer in 1980. In this book, Kidder observed that computer engineers often harbor strong feelings toward their new designs (Kidder, 1981). Likewise, specialist HPC designers are usually passionate about the technology they are building, and territorial about their space in the industry. Moreover, those who are funded by the government have to propose a project, defend their budget, and as in the private sector fight off rivals who are designing competing machines and fighting for the same sources of money. This process becomes part of their norms and reinforces a culture of inward focus and exclusivity.

Although there are exceptions, users of HPC and the people who program them in the Department of Defense (DOD) and government laboratories are nearly all advanced specialists in their fields with many years of experience using HPC to solve difficult problems. In many cases the users and authors of codes are the same group of researchers, forming—because of the rarified nature of the work—something of an exclusive club.

Because of this, HPC systems and their applications have very steep learning curves for novice users who have not spent years programming them. HPC applications often solve a broad range of problems ranging from modeling the universe to designing new aircraft, and users are required to specify hundreds of input variables related to the problem to be solved; mis-specifying even one of these parameters can result in an aborted run or, worse, a completed run with a nonsensical answer. Because HPC system vendors and large federal HPC programs have had the luxury of a large base of experienced users, the systems themselves remain quite difficult to use for outsiders and in general do not reflect many of the lessons in human-computer interaction that make today's cellphones, tablets, and laptop computers relevant and accessible to a wide range of users in our society.

All these factors have limited the adoption of advanced technical computing in emerging areas by nonspecialists, or lower- to middle-end users. A focus on creating more usable HPC systems that utilize modern interfaces and provide more cognitive support to non-HPC specialists would lower the barrier to entry for new application areas, such as HPC-enabled analysis of complex social and behavioral structures and allow for new forms of non-CMOS computing to be experimented with. An emphasis on user productivity with mission-critical application codes could simplify correct specification of problems and execution of advanced simulation applications by nonspecialists, increasing the applicability of tools in existing areas of use to users from new communities.

The United States, as a nation, is at a major inflection point in developing next-generation, high-end, extreme-scale systems. We are moving from computing based on processors that are programmed to follow a predesigned sequence of instructions, to the cognitive computing era based on massive amounts of data and systems evolving into systems that can "learn," such as Amazon's Echo platform that will figure out that you like veggie pizza and mystery novels. Cognitive systems can modify and optimize projections or weigh the value of information based on experience and results. This new approach to computing requires entirely new strategies and skills to maintain U.S. leadership in technology. If we solve the challenges in exascale that include power, energy, memory, storage, concurrency and locality, and resiliency, not only will we advance technology as a discipline, but if humanity chooses to put this technology to good use we will also bolster economic competitiveness and bring unprecedented understanding of the natural world and the cosmos.

The new imperative must be to design for data, not for processor performance. Industry's focus has largely turned to Graphical Processing Unit (GPU) handhelds and consumer products.[8] The rest of the world is going in a different direction than the U.S. HPC community: everyone's iPhone is, in principle, a supercomputer. Consumer demand has given rise to a massive data intensive-based computing industry of generalist users. Notably, Japan's Fugaku is the first system powered by ARM processors—a significant architectural milestone of which a detailed analysis of is beyond the scope of this chapter. Unfortunately, many leaders in the HPC community see the rise of data computing as an either-or question: either we invest in exascale or in data computing. We need to push on both. Thinking back to Joe's class, Thomas Kuhn's landmark *The Structure of Scientific Revolutions* (1962) comes to mind where he introduced the concept of a paradigm shift in scientific research and belief (Kuhn, 1962). A paradigm shift is now desperately needed in supercomputing, because, as things stand today, *meaningful, usable, and reliable exascale computing is still unattainable.*

It's hard to see the future, but economics will no doubt drive change in the HPC community and industry. And it may well be a small startup company working on non-CMOS computing that overthrows the current HPC industry. This scenario would be truly disruptive, where the future may be the forced production of very different types of microprocessors based on volume economics. Indeed, as Neil Thompson and Svenja Spanuth have argued in a recent working paper, the combination of the end of Moore's Law causing notable rising fixed costs for the semiconductor industry and general-purpose universal processors becoming a mature technology, more and more advanced computing users will switch to specialized processors for their needs, leaving HPC as we know it behind (Thompson and Spanuth, 2019).

Adding to the economic aspect, individual government agencies do not have uniform HPC needs for their respective missions. Historically, while the Department of Energy (DOE) and the Department of Defense (DOD) have driven the very high end of HPC—often at hefty price tags of hundreds of millions of dollars per each new machine acquisition—areas such as weather and climate modeling, and basic military vehicle design and ordnance computing needs generally can be carried out using lower end computing such as grids, clouds, or white box clusters. Being its biggest customers, the DOE and DOD have been essentially keeping the HPC industry alive for the past several decades. This worked through the 1980s and perhaps 1990s. But this business model is no longer sustainable.

At the Harvard Business School, the late Clay Christensen wrote extensively about innovation and technology and his work has had almost as much influence on my thinking as Joe's. Christensen came up with the disruptive innovation concept, which describes a process by which a product or service takes root initially in simple applications at the bottom of a market and then relentlessly moves upmarket, eventually displacing established competitors. In Christensen's *The Innovator's' Dilemma* (1997), he examines several cases of disruptive innovation occurring in the disk drive, steel, and mechanical excavator industries, where the large established corporations in each of these respective areas did not anticipate the small disruptor upending their markets and found themselves losing to the new competitors. HPC today is a cutthroat, rapidly consolidating business, and not a high margin one to boot.[9] Competition from upstarts may bury the HPC business.

Christensen wrote more recently about how modern companies' pursuit of profits and myopic focus on return on net assets (RONA), internal rate of return (IRR), and return on capital employed (ROCE) is killing America's ability to generate empowering innovations (Christensen, 2012). The HPC industry is as guilty of this behavior as any other big industry, where, as Christensen has described:

> In the semiconductor industry, for instance, there are almost no companies left in America that fabricate their own products besides Intel [INTC] [*sic*]. Most of them have become "fab-less" semiconductor companies. These companies are even proud of being "fab-less" because their profit as a percent of assets is much

> higher than at Intel. So they outsource the fabrication of the semi-conductors to Taiwan and China. (Christensen, 2012)

The U.S. government is no longer driving innovation in tech. Consumer capitalism is. Silicon Valley's ascent and strong innovation culture was early on dependent on the U.S. government. China's current rapid ascent in technology is strongly supported by its government, which focuses on the long-term view of investment, and not on RONA, IRR, or ROCE. Furthermore, Chinese industry is not bifurcated from its government the way it is here at home. The United States may be facing a sputnik moment if China reaches meaningful exascale computing first; China has publicly announced plans to achieve exaFLOP speeds by 2021.

The United States, for the first time in decades, faces technological uncertainty from a formidable economic and growing military competitor. The logical positivists shouted a warning call to the Western world when they struck out against fascism and called for ground truth on scientific inquiry. Although we already live in a near completely globalized world in terms of industrial supply chain and manufacturing, China's stunning ascent in HPC should serve as another kind of warning to Western nations: that we are entering a new era of global balance of power that is very different than that which characterized the Cold War.

The U.S. government has had a great deal of influence on both scientific and technological discoveries and developments for the past seven decades. Yet, the government has gradually been shifting away from funding big technological projects over the last thirty years. Consider the Human Genome Project that was completed only when J. Craig Venter stepped in with private money, or the current privatization of space travel, or the quest to end malaria that has largely been funded by the Gates Foundation. Wealthy patrons are taking up the mantle of sponsoring some of these big projects where the government has lagged. US HPC could see the same fate, which might allow for US supercomputing to get back on track toward convergence of classical and data computing and solve the challenges to exascale.

In writing this chapter, I do not, nor does anyone I know or work with, claim to offer a single correct solution to getting US HPC on track because there isn't one all-encompassing answer to this multifaceted problem. What is clear is that HPC, as a technological case study, demonstrates how hard it is for human beings to *foresee* and *prepare* for major technology changes. And as human beings in the thick of the HPC business, it's easy to miss all the technological and economic changes that are going on right in front of us. Compared to Galileo's telescope, technology today is very complex in terms of components and their global supply chain, politics, personalities, culture, corporate and national interests, and innovation drivers. Market forces are as important in shaping the direction of technology as are all these other factors covered in this chapter.

Given that HPC as a case study has been affected by all these forces, technology does stand on its own apart from science as Joe has shown us in his work, and not merely technology, but I would argue that increasingly, given its pervasiveness, it is computing that is defining the human condition today. If technology is humanity at work, then computing is humanity running a marathon given the rate of change in this field. Computing today looks very different and substantively is very different than what it was forty years ago. And it will likely be unrecognizable by early twenty-first-century humans if we could see what computing will look like forty years from now.

One of the valuable lessons about studying with a philosopher like Joe is that sometimes it pays to look at a problem from a completely different view. So from another perspective, perhaps "saving" HPC as an industry or enterprise is not the right answer. Usable exascale may simply not be reachable. We know that machoflops don't matter. General-purpose computing may be too mature of a technology to advance rapidly anymore, as the slowing down in their improvements in performance-per-dollar seems to be showing, and specialized processors have become the norm (Thompson and Spanuth, 2019). Computer platforms and systems are changing, as we've seen, and people will ultimately adapt to new technology offerings.

Long ago, Joe spurred an effort to see and analyze technology as something that people do and how they go about doing it (Pitt, 2000). Joe's work has influenced me to approach the problem of how to move advanced computing forward in a holistic way that includes considering humans and technology both as part of the solution, and gives me hope that I am making the most informed decisions when I am contributing to public sector decision-making—or "making the sausage" in policy speak—that involves nine-figure budgets. No one can ultimately see the future of technology, but Joe's writings on the philosophy of technology have helped the technological horizon seem a little less murky, and for that knowledge, I am grateful to him.

NOTES

1. The views expressed are the author's own and do not represent those of the US Government.

2. See Top500.org. Since 1993 the Top500 project ranks and describes the 500 most powerful HPC systems in the world. The list is updated twice per year.

3. HPC is often viewed as a measure of a country's economic competitiveness. To emphasize this, the US Council on Competitiveness coined the phrase, "To

outcompete is to outcompute" several years ago. See https://www.compete.org/programs/compete-innovation.

4. LINPACK is a software library for performing numerical linear algebra on digital computers by running a program that solves a system of linear equations. It was written in the 1970s.

5. Countries voluntarily participate in this ranking: there exist many fast supercomputers in classified military settings around the world that do not appear on the Top500 list.

6. I have often wondered what the HPC industry would look like and if machoflops would even exist as a term if the supercomputing business were dominated by women instead of men. Would the Top500 list exist at all? Or would computing architecture look very different than it does now?

7. HPC culture has some similar characteristics to that of high energy physicists as described by Sharon Traweek in *Beamtimes and Lifetimes: The World of High Energy Physics*, Cambridge: Harvard University Press, 1992.

8. In contrast to a Central Processing Unit (CPU), which Intel makes, for example. CPUs and GPUs are physically not that dissimilar. They are both composed of hundreds of millions of transistors, and can process thousands of operations per second. CPUs are often colloquially referred to as a computer's brain, handling an astonishingly wide variety of types of problems and forming the basis of general-purpose computing. A GPU is a specialized type of microprocessor optimized to display graphics and does very specific computational tasks. It runs at a lower clock speed than a CPU but has many times the number of processing cores. GPUs are popular among gamers and are used increasingly to perform data computing.

9. In contrast, public cloud building such as that done by Amazon Web Services is a high margin business.

REFERENCES

Byrne, D. M., S. D. Oliner, and D. E. Sichel. 2013. "Is the Information Technology Revolution Over?" *International Productivity Monitor*. 25: 20–36.

Christensen, Clayton M. 2012. "A Capitalist's Dilemma, Whoever Wins on Tuesday." *The New York Times*, November 3 [Accessed February 15, 2019].

Dennard, Robert H., Fritz Gaensslen, Hwa-Nien Yu, Leo Rideout, Ernest Bassous, and Andre LeBlanc. 1974. "Design of Ion-Implanted MOSFET's with Very Small Physical Dimensions." *IEEE Journal of Solid-State Circuits*. 9(4): 256–68.

Haraway, Donna J. 1991. *Simians, Cyborgs, and Women: The Reinvention of Nature*. New York: Routledge.

Hempel, Carl G. 1965. *Aspects of Scientific Explanation: And Other Essays in the Philosophy of Science*. New York: Free Press.

Kidder, Tracy. 1981. *The Soul of a New Machine*. New York: Avon Books.

Kuhn, Thomas. 1962. *The Structure of Scientific Revolutions*. Chicago: The University of Chicago Press.

MacKenzie, Donald. 1991. "The Influence of the Los Alamos and Livermore National Laboratories on the Development of Supercomputing." *IEEE Annals of the History of Computing*. 13(2): 179–201.

Marsden, Rhodri. 2013. "Mickey Is Watching You: Does Disney's New 'Magic Band' Infringe on Consumer Freedoms?" *The Independent*, February 20, 2013 [Accessed February 9, 2019].

Mody, Cyrus C. M. 2017. "Moore's Regula." In *Spaces for the Future: A Companion to the Philosophy of Technology*, edited by Joseph C. Pitt and Ashley Shew, 238–47. New York: Routledge.

Pitt, Joseph C. 1992. *Galileo, Human Knowledge, and the Book of Nature: Method Replaces Metaphysics*. Dordrecht: Springer Science + Business Media.

Pitt, Joseph C. 2000. *Thinking about Technology: Foundations of the Philosophy of Technology*. New York: Seven Bridges Press.

Pitt, Joseph C. 2017. "Transcendence in Space." In *Spaces for the Future: A Companion to the Philosophy of Technology*, edited by Joseph C. Pitt and Ashley Shew, 340–45. New York: Routledge.

Popper, Karl. 1968. *The Logic of Scientific Discovery*. New York: Harper & Row.

Popper, Karl. 1994. *The Open Society and Its Enemies*. Princeton: Princeton University Press.

Rossiter, Margaret. 1982. *Women Scientists in America: Struggles and Strategies to 1940*. Baltimore, MD: Johns Hopkins University Press.

Snow, C. P. 1962. *The Two Cultures and the Scientific Revolution*. New York: Cambridge University Press.

Thompson, Neil C. and Svenja Spanuth. 2019. The Decline of Computers as a General Purpose Technology: Why Deep Learning and the End of Moore's Law Are Fragmenting Computing. Working Paper at https://ssrn.com/abstract=3287769 [Accessed April 2, 2019].

Traweek, Sharon. 1992. *Beamtimes and Lifetimes: The World of High Energy Physics*. Cambridge: Harvard University Press.

Chapter 2

The Applicability of Copyright to Synthetic Biology

The Intersection of Technology and the Law

Ronald Laymon

Professor Pitt has forcefully stated what the fundamental role of philosophy should be:

> Philosophy, it is suggested, should be about mankind interacting with the world, which is, on my account, *the nature of technology*. The role of philosophy should be to help us accomplish those interactions in a thoughtful and productive manner. *The philosopher should be seen as part of a team of individuals seeking to accomplish something*—she is a critical facilitator—Socrates reborn. . . . I urge that we first see the aim of philosophy to be assisting humankind to make their way in the world. (Pitt 2016, 83, emphasis added)

Accordingly, and with his usual infectious enthusiasm, he issued the following challenge:

> In [my] view we [should] cease to think of ourselves as metaphysicians or epistemologists, but as philosophers who *can help identify metaphysical or epistemological issues* as they arise while we do our work together with engineers, geologists, social planners, etc. The idea here is not to see the traditional areas of philosophy as separate areas of research in their own right, but as problem areas that need to be identified and dealt with in specific contexts. (Pitt 2016, 87, emphasis added)

While overly optimistic in the ordinary course of academic life, it is still, without doubt, a worthy *aspirational* goal. And when the stars and planets happen to propitiously align, we should be prepared to act. Promising opportunities for such alignment and consequent execution on Professor Pitt's challenge arise, as he has emphasized, with regard to matters of law.

> So *philosophy of law* can be taken as part of *philosophy of technology*. When we undertake a legal action we engage in activities which affect the lives and fortunes of many others. The philosopher's role is to assist in *ferreting out the implications* of this or that legal move or attempt to change the legal system. We should be working with lawyers, judges, plaintiffs, and legislators to help determine the best path of action. (Pitt 2016, 88, emphasis added)

My contribution toward this aspirational goal and associated "ferreting out of implications" deals with a matter of proposed legal reform in the area of intellectual property that's squarely at the intersection of law, technology, and the public interest. Here's the question: *Should copyright law be made applicable to a subset of biological research and development known as synthetic biology?*

Why this question? This is currently a contentious and challenging issue involving working scientists, law school academics, practicing attorneys, and various public interest and trade groups. Also implicated is the ongoing tension between advocates of "open source" pharmaceutical research and development, and "Big Pharma's" reliance on patent protection.[1] This is thus *clearly a matter of significant social policy*. And accordingly, exactly the sort of issue on which professor Pitt has focused the need for attention. Moreover, reviewing the intricacies here will provide an entry into the ways and means attorneys and the courts use to deal with matters both scientific and technological, especially where there is stretching of the existing doctrinal categories to the breaking point in order to accommodate new forms of scientific and technological development.

THE ESSENTIALS OF PATENT AND COPYRIGHT LAW

By way of providing some necessary background, we'll need to briefly review the basic principles of patent and copyright law. As an initial matter, these principles derive their origin and legitimacy from the patents and copyright clause of the U.S. Constitution (Article I, Section 8, Clause 8), which empowers Congress "to promote the progress of science and useful arts, by securing for limited times to authors and inventors the exclusive right to their respective writings and discoveries." In short, and importantly so, patent and copyright law are at bottom matters of a Constitutionally mandated *public policy*, namely, "to promote the progress of science and useful arts."

Patent Law

Section 101 of the U.S. Patent Act specifies that any person who "invents or discovers any new and useful process, machine, manufacture, or composition of matter, or any new and useful improvement thereof, may obtain a patent." In its explication of this provision, the Supreme Court has long held that "laws of nature, natural phenomena, and abstract ideas" are exceptions and thus not patent eligible (*Association for Molecular Pathology* 2013, 10–11). Once a patent application passes this § 101 *threshold test of eligibility*, it is then subjected to the requirements of being novel (§ 102) and nonobvious (§ 103).

Applications for a patent must demonstrate, to the satisfaction of the U.S. Patent and Trademark Office (USPTO), that the statutory requirements have been satisfied. This is no easy matter, especially since the USPTO application manual is over 1,000 pages long! Not surprisingly, the application process is complicated, expensive, and slow. With respect to synthetic biology, recent successful patent applications have varied from 25 to 250 long and grueling pages.[2] More details on the requirements for such an application will be discussed later. And while a patent is nominally good for twenty years, there is no guarantee that it will not be legally challenged as infringing another patent. Since the USPTO is an administrative agency, its decisions can be legally challenged.

Copyright Law

The basic copyright requirements of relevance here are as follows:

> Copyright protection subsists, in accordance with this title, in *original works of authorship fixed in any tangible medium of expression*, now known or later developed, from which they can be perceived, reproduced, or otherwise communicated, either directly or with the aid of a machine or device. (17 U.S. Code § 102(a), emphasis added)

> *In no case does copyright protection for an original work of authorship extend to* any idea, procedure, *process*, system, *method of operation*, concept, principle, or discovery, regardless of the form in which it is described, explained, illustrated, or embodied in such work. (17 U.S. Code § 102(b), emphasis added)

With respect to the originality requirement of § 102(a), the Supreme Court has determined that "originality requires independent creation plus a modicum of creativity" (*Feist Publications* 1991, 346). So not much by way of creativity is required. Section 102(a) also provides a non-exhaustive list of the many types of authorship that are copyrightable. Notably, computer programs do not appear on the list. But the courts have determined, as a matter of accommodation, that they're to be considered as a form of "literary work" and thus eligible for copyright protection.[3] Obviously a stretch, but, while the Copyright Act does not define "computer," it does define a "computer program" as "a set of statements or instructions to be used directly or indirectly in a computer in order to bring about a certain result" (17 U.S.C. § 101).

With regard to § 102(b), it's important to note that while "any new and useful process" is eligible for patent protection, § 102(b) expressly excludes copyright protection from extending to any "idea, procedure, *process*, system, *method of operation*, concept, principle, or discovery . . . embodied" in copyrighted material. The role of the expression "idea" in § 102(b) is not immediately evident. As used here, it's a term of art which in its legal application occurs as part of a central doctrinal distinction, namely, that between the "expression" for which copyright protection is sought and the "idea" that the expression instantiates. The distinction has a long common law history and is best introduced by way of example. Consider *The Thin Man* by Dashiell Hammett. In the context of copyright law, that work is to be understood as the *expression* of the general *idea* of a murder mystery novel. The idea-expression distinction then demarcates what may be copyrighted, namely, the particular *expression*, from that which cannot, the general *idea*.

Another key concept in the law of copyright is that of *merger*. Putting aside certain niceties until later, the basic idea is that if the expression at issue is the only way to express the associated idea, then that expression cannot be copyright protected—assuming that the idea is not something that can or should be copyright protected. Otherwise, the holder of a copyright on the expression would, in effect, gain ownership over the idea because no one else could use and thus express the idea other than by violating that copyright. The doctrine owes its origin to the case of *Baker v. Selden*, where the court held that while a book describing a bookkeeping system is worthy of copyright protection, the underlying method described is not. Thus, the author of the book could not assert copyright infringement against a statement or equivalent paraphrase of the method, or against the employment of the method, as in the case of a producer of ledger sheets specifically designed for the use of the method. Merger is thus, in essence, a prohibition against the granting of copyright to expressions at high levels of abstraction—which would violate the Constitutional requirement that such an "exclusive right" may be granted only "to promote the progress of science and useful arts."

While the patent application process is arduous, slow, expensive, and often uncertain as to effectiveness, with regard to copyright it is no longer even necessary to mark an original work of authorship as "copyright." Nor need registration be made with the U.S. Copyright Office. It is enough simply to fix the authorship in some "tangible medium of expression." Registration, however, does establish prima facie satisfaction of copyright requirements and thus places the burden of proof on the opposing party contesting the copyright. Registration is also required in order to initiate an infringement action. And since it's not required to establish copyright, registration requirements are minimal and nothing like those required for patent application. In essence, all that needs to be done is to provide a copy of the "authorship," what had been earlier referred to in the Constitution as the "writing," along with some indication of originality and a "modicum" of creativity. Finally, copyright is good for ninety-five years (a first, term of twenty-eight years plus a renewal term of sixty-seven years).

Summary and Comparison of Patent and Copyright Law

In sum, patent protection is expensive but offers "thick" protection of the entirety of the invention, including any process or method of operation involved. By contrast, copyright is vastly less expensive but provides only "thin" protection of the "writing." In addition to expense and level of protection, there is another important difference, namely, computer programs are entitled to *both* copyright and patent protection, whereas biological discoveries and inventions are entitled only to patent protection. The question raised then is, why not extend the copyright benefits afforded to computer programs to the DNA codings, the "writings" of synthetic biology?

The Argument to Provide Copyright Protection for the DNA Specification Sequences of Synthetics Biology

The following captures the essence of an argument that was recently presented to the US Copyright Office for registration of the DNA sequence for a synthetically created fluorescent protein (the Prancer DNA Sequence), and thus, in effect, to extend copyright protection to this instance of synthetic biology:[4]

(1) Computer programs are eligible for copyright protection.
(2) The DNA specification sequences of synthetic biology are analogous to computer programs. Both are "writings" that are "original works of authorship" that have been "fixed" in "a tangible medium."
(3) The copyright legal regime is well developed for the consideration of computer programming and can accordingly accommodate the entry of synthetic biology.
(4) Therefore, copyright protection should be extended to the DNA specification sequences of synthetic biology.

The principal and motivating benefit of the proposed expansion of copyright protection is the simplification achieved and resulting reduction in expense, since *the only substantial registration requirement is the specification of the DNA sequence*. This stands in stark contrast to what's required in a patent application. But as already noted, in exchange for such simplification comes intellectual property protection that is, at least on its face, less extensive.

The Copyright Office, however, was not moved by the argument and declined the request. The crux of the dispute was, not surprisingly, the claimed analogy between computer programs and the DNA specification sequences of synthetic biology. The registration applicant for its part claimed that because of the analogy:

> [T]here is nothing in copyright law that would justify treating a set of instructions directed towards a computer any differently than a set of instructions directed towards some other machine capable of receiving and acting upon the instructions, including a biological machine such as a recombinant microorganism. (Holman 2016, 122)

The Copyright Office disagreed and bluntly dismissed the relevance of the claimed analogy:

> The 231 codons that make up the Prancer DNA Sequence are not statements or instructions that are used directly or indirectly in a computer in order to bring about a certain result. While an organism may be analogized to a machine, it clearly is not one, and as a result falls outside of the category enumerated in the statute. (Holman 2016, 122. Internal quotation omitted)

In other words, the DNA specifications to be used in a "recombinant organism" do not fall in the § 102(a) category of "literary work" because the "organism" is not a computer. The Office's rejection is puzzling because, as already noted, the Copyright Act does not define what constitutes a computer. But the Copyright Office, in accordance with its administrative powers, did promulgate the following definition:

> For purposes of copyright registration, a "computer" is defined as a programmable *electronic* device that can store, retrieve, and process data that is input by a user through a user interface, and is capable of providing output through a display screen or other external output device, such as a printer. "Computers" include mainframes, desktops, laptops, tablets, and smart phones. (Copyright Office 2017, § 721.2, emphasis added)

Thus the Copyright Office was correct in its dismissal of the claim that recombinant organisms were computers, but only insofar as its own administration definition was taken to be definitive—which as a matter of administrative law it is not.[5] Even so, the Office refused to budge and suggested that the registrant appeal to federal court. Ironically because of the expense involved, the applicant decided not to pursue an appeal.

Before considering what would constitute an effective basis for such an appeal, we'll need to briefly consider a supplemental argument that the Copyright Office offered in support of its rejection of the copyright request. The Office claimed that § 102(b) "precluded" the Prancer DNA Sequence from copyright protection. As noted above, the provision provides that "[i]n no case does copyright protection for an original work of authorship extend to any . . . process [or] method of operation." The general sense of the provision seems clear enough, and was further elaborated in the legislative history of the provision:

> Some concern has been expressed lest copyright in computer programs should extend protection to the methodology or processes adopted by the programmer, rather than merely to the "writing" expressing his ideas. Section 102(b) is intended, among other things, to make clear that the expression adopted by the programmer is the copyrightable element in a computer program, and that the actual processes or methods embodied in the program are not within the scope of the copyright law. (H.R. Report 1976, 57)

Taken on its face, the relevance of § 102(b) is that even if the Prancer DNA Sequence were copyrightable, that copyright protection would not extend to the "actual processes or methods embodied" in that sequence. Accordingly, those processes or methods are not copyrightable. There is, of course, the question of what exactly constitutes such embodied processes or methods, but more on this below. In any case, this is not how the Copyright Office understood and explained the applicability of § 102(b), stating instead that the provision "precluded" the Prancer DNA Sequence from copyright protection because:

> The Prancer DNA Sequence is genetic formula for a biological system. It does not describe, explain, or illustrate anything except the genetic markers that comprise this biological organism. Therefore, there is no copyrightable expression, but rather the claim simply records the formula for this biological system or process. (Holman 2016, 122)

The relevance of the comment is, however, opaque at best, and less charitably read, decidedly irrelevant because § 102(b) applies in cases where there is a copyrightable expression and restricts that protection to the expression itself and not to any process or method of operation that is embodied in that expression. By contrast, the Office's explanatory comment deals with the question of whether the Prancer DNA Sequence is copyrightable in the first place—and as such is no more than a further elaboration of its rejection of the claimed analogy between the sequence and a computer program. In its defense, the Office may have thought, though it did not expressly say, that § 102(b) serves to strip away the *relevance* of the emergent functional aspects of the DNA sequence—the "embodied" processes and methods of operation—for the copyrightability of synthetic DNA once incorporated into a recombinant organism. And once this relevance is stripped away, all that's left is a "genetic formula for a biological system," which, while perhaps patentable, is *not* copyrightable.

Before any headway can be made on resolving the dispute regarding the copyrightability of the Prancer DNA Sequence, and more generally of the DNA sequences of synthetic biology, we need a better grip on the specifics regarding in what sense such sequences are to be construed as "a set of instructions" controlling the operation of "a recombinant microorganism." And given these specifics, what follows *as a matter of law* regarding copyright protection. In other words, what is the relevance and strength of the claimed analogy between the DNA coding sequences of synthetic biology and computer programming?

SYNTHETIC BIOLOGY AND THE ANALOGY WITH COMPUTER PROGRAMMING

The underlying mantra that motivates synthetic biology is to mimic the design of electronic circuits by creating a repertoire of well-described, modular biological components with stable performance characteristics that can be combined and then expressed in a host recombinant microorganism, known as the "chassis"—typically some variant of *Saccharomyces cerevisiae* yeast or *Escherichia coli*—thereby creating biological analogues to electronic circuits.[6]

The *lac* operon, variants of which are now available in synthetic form, provides a paradigmatic illustration of the mantra in action. Once incorporated in a compatible chassis, the *lac* operon augments the input-output repertoire of the chassis. Thus, for example, in its natural form and environment, it controls how the internal digestive capabilities of E. coli are triggered to respond appropriately in a glucose-rich environment which is maximally nutritious, or in an alternative lactose environment which is not as nutritious.[7] See figure 2.1, which depicts the general methodology of synthetic biology and, as an illustration of how its application would operate in synthetic form, the performance characteristics of the *lac* operon, including the production of ß-galactosidase in lactose environments.

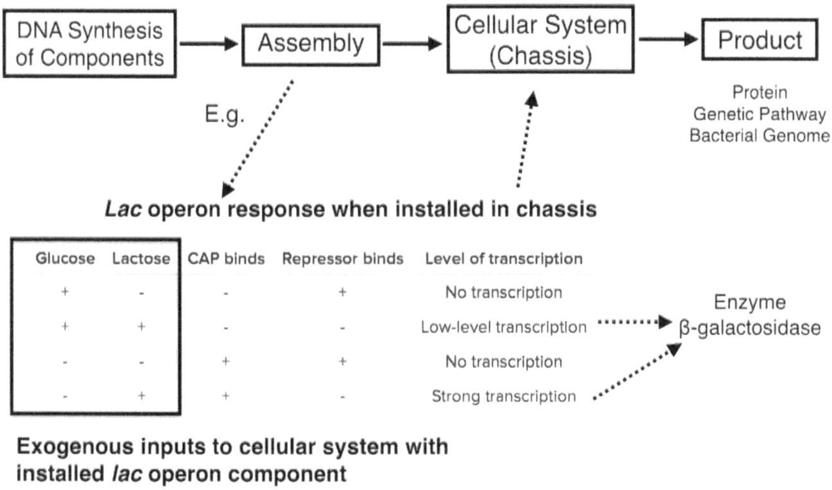

Figure 2.1 The General Methodology of Synthetic Biology, and the *Lac* Operon as an Example of Emergent Functionality. *Source*: Created by the author.

The question posed by our exemplar is, what *exactly* is the legally relevant analogy to computer programming supposed to be? In seeking the precise nature of the claimed analogy, it's instructive to start not with a modern-day computer but rather with an early precursor, namely, ENIAC. This is because ENIAC did not have a stored program. Its components (accumulators, multiplier, divider, square rooter, cycling and control units, input-output devices, and stored function tables) had to be *hard wired* together in order to effectuate the particular desired computation.[8] The specific and special purpose hard wiring was thus the "program." Similarly, once incorporated in its chassis, synthetic DNA (such as the *lac* operon) acts analogously to introduce "hard wired" connections with the existing transcriptional and input-output capabilities of the chassis.

Whatever analogy exists, therefore, between synthetic DNA sequences and computer programming must have as its foundation the fundamental, underlying analogy between the *composite* of incorporated DNA and cellular chassis and the *composite* of ENIAC augmented by its special purpose, hard-wired programming. The proposal to make copyright applicable to the DNA sequences of synthetic biology *severs this connection* between the DNA and its chassis in the sense that the "writing" to be copyrighted is the DNA sequence considered in *isolation* from the *specific* functional consequents of the incorporation of the DNA so described in a compatible chassis. The claim then is that the DNA sequence *alone* serves as the "writing" to be copyright protected, and because of that, no specifics regarding the functional consequences of the DNA once incorporated in any chassis are required for copyrightability. This is the fundamental, simplifying advantage of the proposal. Moreover, because of the prohibition imposed by § 102(b), the modular performance characteristics of the composite system of DNA and chassis are not, in any case, copyrightable. By contrast, a patent application must include both the DNA component and the performance characteristics once installed in its chassis, and much more besides.

A FUNDAMENTAL DISANALOGY BETWEEN COMPUTER SOURCE CODE AND THE DNA SEQUENCES OF SYNTHETIC BIOLOGY

In support of their focus on the DNA sequence alone, proponents of copyright for synthetic DNA often speak as if the *relevant and controlling analogy* is between the specification of a DNA sequence and computer *source code* programming as installed and executed on contemporary computers.[9] There is, however, a fundamental *disanalogy* between the DNA sequences of synthetic biology and computer source code programming. Keeping ENIAC in mind, the disanalogy is readily apparent:

> Source Code Program: The functionality and modularity is evident *as a matter of logic* given the syntax and semantics of the underlying language.
>
> The DNA Coding for Synthetic Biology: The functionality and modularity is *invisible* and not evident as a matter of syntax and semantics but depends on the microbiological specifics of the interaction of the DNA with the transcription and translation capabilities of the intended chassis.

And here we note that the wiring connections used to program ENIAC are similarly "invisible" as to their functional effect, which can be retrieved only by an analysis of the physical processes initiated and facilitated by those connections. So too for modern-day stored programs written in *object* (machine) code. By contrast, the functionality and modular structure of contemporary source code are clear *on its face* precisely because it has been so designed and implemented.

Because of the difference between object and source code with regard to the visibility of resultant functionality, the U.S. Copyright Office applies the Rule of Doubt when registrations are made for an object as opposed to for a source code. In such cases, the Office acknowledges that a registration request has been filed but states that it cannot confirm the presence of copyrightable authorship because it has no effective basis (given the functional invisibility of object code) to determine whether the originality and creativity requirements are satisfied.[10] Accordingly, even putting aside its general objections to the copyrightability of DNA sequences, the Copyright Office would similarly be obliged to apply the Rule of Doubt to any registration of a DNA sequence. This is because of the invisibility of the functional consequences of the DNA sequence *when viewed in isolation.*

This disanalogy between source code programming and DNA has been overlooked by proponents of extending copyright protection to synthetic biology. This is why the claim is that, for purposes of copyright, only the "writing" of the DNA sequence needs to be considered. And as already noted, because of its simplicity, this narrow focus is the principal charm and apparent advantage of the copyright proposal. If, however, a specification of the performance characteristics of the combination of DNA coding and chassis is not to be required in copyright registration, or more generally for copyrightability, then what's left is simply *the DNA sequence as a list of instructions for the construction of a physical thing* to be incorporated *with unspecified functional effect* in some cellular chassis. Accordingly, and here's the rub, such a recipe for construction is not copyrightable for the simple reason that it does not fit within any of the specified categories of copyrightability. While the DNA at issue might be claimed to be a "useful article" insofar as it serves to produce a useful *though unspecified* result, that doesn't help because "useful articles" are copyrightable only with respect to their separable *artistic* features where those features "can be identified separately from, and are capable of existing independently of, the utilitarian aspect of the article" (Copyright Act § 101). Clearly, there are no such legally relevant artistic features here.[11]

The Copyright Office may have had this sort of analysis in mind, albeit in the context of § 102(b), when it claimed that the Prancer DNA Sequence was "precluded" from copyright protection by that provision because the sequence "does not describe, explain, or illustrate anything except the genetic markers that comprise this biological organism." If so, then the Copyright Office apparently thought that since the "embodied" functionality of synthetic DNA was not copyrightable because of § 102(b), this meant that *all that remained of relevance* for a determination of copyrightability was the DNA sequence as a set of non-copyrightable instructions for the creation of a "useful article." As will be shown below, however, the Copyright Office would have been mistaken in so restricting the relevance for copyrightability of the "embodied" functionality.

In any case, if synthetic DNA is to avoid a determination of being no more than a useful article, then the copyright *credentials* for synthetic DNA must somehow *expressly* include not only the DNA sequence itself, but also the functional effect—the "embodied" process or method of operation—of the DNA once installed in its chassis. For a specific instance of the functional

```
>BBa_K389050 Part-only sequence (1600 bp)
tggtgcaaaacctttcgcggtatggcatgatagcgcctactagagaaagaggagaaatactagatggtgaatgtgaaaccagtaacgttatacgatgtcg
cagagtatgccggtgtctcttatcagaccgtttcccgcgtggtgaaccaggccagccacgtttctgcgaaaacgcgggaaaaagtggaagcggcgatggc
ggagctgaattacattcccaaccgcgtggcacaacaactggcgggcaaacagtcgttgctgattggcgttgccacctccagtctggccctgcacgcgccg
tcgcaaattgtcgcggcgattaaatctcgcgccgatcaactgggtgccagcgtggtggtgtcgatggtagaacgaagcggcgtcgaagcctgtaaagcgg
cggtgcacaatcttctcgcgcaacgcgtcagtgggctgatcattaactatccgctggatgaccaggatgccattgctgtggaagctgcctgcactaatgt
tccggcgttatttcttgatgtctctgaccagacacccatcaacagtattattttctcccatgaagacggtacgcgactgggcgtggagcatctggtcgca
ttgggtcaccagcaaatcgcgctgttagcgggcccattaagttctgtctcggcgcgtctgcgtctggctggctggcataaatatctcactcgcaatcaaa
ttcagccgatagcggaacgggaaggcgactggagtgccatgtccggttttcaacaaaccatgcaaatgctgaatgagggcatcgttcccactgcgatgct
ggttgccaacgatcagatggcgctgggcgcaatgcgcgccattaccgagtccgggctgcgcgttggtgcggatatctcggtagtgggatacgacgatacc
gaagacagctcatgttatatcccgccgttaaccaccatcaaacaggattttcgcctgctggggcaaaccagcgtggaccgcttgctgcaactctctcagg
gccaggcggtgaagggcaatcagctgttgcccgtctcactggtgaaaagaaaaaccaccctggcgcccaatacgcaaaccgcctctccccgcgcgttggc
cgattcattaatgcagctggcacgacaggtttcccgactggaaagcgggcaggctgcaaacgacgaaaactacgctttagtagcttaataactctgatag
tgctagtgtagatctctactagagccaggcatcaaataaaacgaaaggctcagtcgaaagactgggcctttcgttttatctgttgtttgtcggtgaacgc
tctctactagagccaggcatcaaataaaacgaaaggctcagtcgaaagactgggcctttcgttttatctgttgtttgtcggtgaacgctctctactagag
caatacgcaaaccgcctctccccgcgcgttggccgattcattaatgcagctggcacgacaggtttcccgactggaaagcgggcagtgagcgcaacgcaat
taatgtgagttagctcactcattaggcaccccaggctttacactttatgcttccggctcgtatgttgtgtggaattgtgagcggataacaatttcacaca
```

Figure 2.2 The DNA Sequence for a Synthetic Variant of the *Lac* Operon. *Source*: Jonas Aretz from the Bielefeld Germany team that participated in iGEM 2010/Registry of Standard Biological Parts hosted by the iGEM Foundation.

invisibility of DNA when viewed in isolation, see figure 2.2, which displays the DNA sequence of a synthetic variant of the *lac* operon.[12]

What's required to pierce this invisibility—and to thereby reveal the underlying, functional aspects of the inserted DNA—is the correspondence between the DNA segments and the modular functional "parts" of the DNA *once it's incorporated into the host chassis*. So, for example, see figure 2.3, which depicts the modular functional parts of a synthetic variant of the *lac* operon once installed in its intended chassis.[13]

This correspondence between the synthetic DNA and its emergent functional consequences is, as discussed earlier, the result of the deliberate design process adopted in synthetic biology, namely, to create a repertoire of modular biological components that can be combined and then expressed in a host recombinant microorganism. These components are constructed and assembled in the hopeful expectation that their modular and functional effect will be as desired once the completed synthetic DNA is incorporated into the target chassis. There is, however, no guarantee that these expectations will be satisfied. Given the complexities and analog processes involved in the underlying microbiological dynamics, the theoretical models employed in the design process are irredeemably idealized and, consequently, at best only approximate. Thus, at a minimum, experimentation and subsequent fine-tuning are required to achieve the desired functionality.[14]

It should now be apparent that the claimed analogy between computer programming and the DNA sequences of synthetic biology *does not stand*

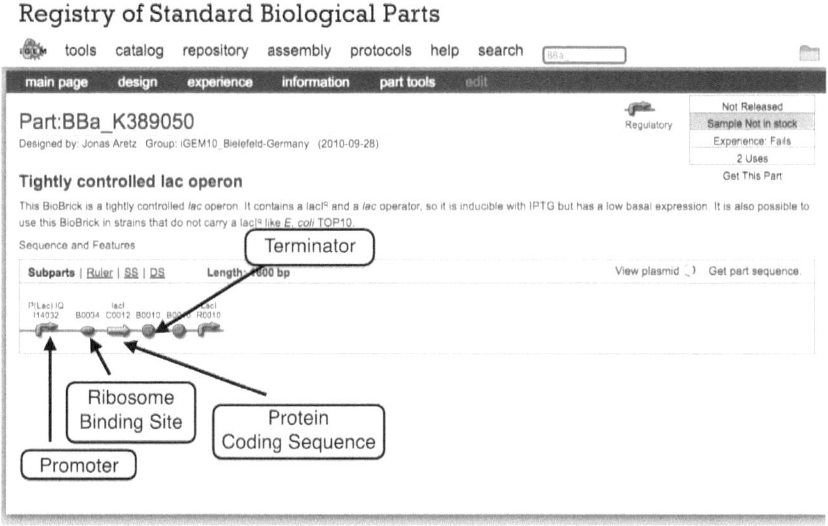

Figure 2.3 The Emergent Modular Functionality of a Synthetic Variant of the *Lac* Operon. *Source*: Adapted from Jonas Aretz from the Bielefeld Germany team that participated in iGEM 2010/Registry of Standard Biological Parts hosted by the iGEM Foundation.

on its own. Because of the functional invisibility of a DNA sequence, the analogy, in order to be truly revealing, must include as an essential ingredient the *emergent* modular and functional consequences of the synthetic DNA *once it's incorporated into a compatible host chassis.* Thus, the relevant analogy is not between the specification of a DNA sequence and computer source code but rather between the *composite* of DNA (understood as hard-wired programming) and chassis, and the *composite* of hard-wired computer code and the computer in which that code is implemented—where the closest such computer analogy is with ENIAC and its hard-wired programming. The corollary here—as will be further elaborated below—is that any claim of copyrightability for a *particular* synthetic DNA sequence must include an acknowledgment of the *specific* chassis in which the sequence is to be incorporated as well as the functional consequences of such incorporation.

With this understanding of the analogy, the problem then for the copyrightability of synthetic DNA is twofold. First, to *justify* as a matter of law why the "embodied" modular functionality of a synthetic DNA sequence should be incorporated as a necessary ingredient of what is required to elevate synthetic DNA from being merely a "useful article" into something that is an instance of authorship that qualifies for copyright protection. Second, to determine how, as a procedural matter, the incorporation is to be effectuated.

With regard to justification, the crucial point to be made is that the analogy with ENIAC is better focused and consequently more determinative than one made with a contemporary computer with stored source code programming. This is because the ENIAC analogy draws attention to the way in which the emergence of the "embodied" functionality of synthetic DNA closely parallels the emergence of the functionality embodied by ENIAC's hard-wired programming—where ENIAC is clearly a "computer" even by the standards of the Copyright Office. In both cases the "invisibility" of the embodied functionality is pierced once the chassis and the ENIAC analogue are brought into play. Based on the historical use of analogy by Congress and the courts, the ENIAC analogy should be sufficient to justify, at least on a prima facie basis, the entry of synthetic DNA into the category of a copyrightable "computer program."[15]

With regard to the procedural question, while there is the possibility of a favorable judicial determination (as in an appeal of rejection by the Copyright Office), the most straightforward way to fully effectuate the necessary incorporation would be by the statutory creation of a sui generis version of copyright law for synthetic biology.[16] In addition to the incorporation of the emergent functionality of synthetic DNA, such a provision should require that the identification of the intended chassis, as well as the emergent functional properties of the DNA once incorporated, be included as necessary ingredients in the initial copyright registration. Otherwise these identifications would have to be determined post hoc in litigation with all the attendant uncertainty and expense involved. Similarly, to avoid such uncertainty and expense, merely fixing the DNA authorship in some "tangible medium of expression" should be deemed insufficient for copyright protection, and registration should be required in all cases.

The ENIAC analogy, however, only goes so far, and is at best prima facie, that is, subject to relevant objection. This is because even assuming incorporation of the "embodied" functionality as a requirement for copyrightability, it would still have to be shown as a matter of public policy that a *coherent, workable copyright system* would result. In particular, any such incorporation of the "embodied" modular functionality must allow for the consistent application of the central doctrinal categories of copyright law, namely, *expression*, *idea*, and *merger*. To unpack the conceptual and legal challenges involved in demonstrating such consistency and applicability, it will be necessary to make a brief detour into the realm of copyright law as applied to computer programming, the other side of the underlying analogy that has been claimed to justify extending copyright coverage to the DNA sequences of synthetic biology.

Merger, Section 102(b) and Computer Programming

One thing to be clear about regarding § 102(b) is that just because a computer program embodies a process, system, or method of operation, it does not mean that it is not copyrightable. That's obviously the case, otherwise no computer program would be copyrightable because, as noted by the Court of Appeals for the Federal Circuit, all "computer programs are by definition functional—they are all designed to accomplish some task," and thereby exemplify a "method of operation" (*Oracle America* 2014, 1367). I mention this apparently straightforward bit of elementary logic because at least one court was tempted into adopting the converse of § 102(b), or was understood to have done so in an appellate reversal, in order to avoid (a variant of) the following problem.[17]

Assume that computer program *P* is copyrightable, and that pursuant to § 102(b) the process, system, or method of operation (henceforth "process") embodied in *P* is not copyrightable. But insofar as *P* is the only way to effectuate the process, the copyright holder thereby achieves the functional equivalent of copyright protection for that process. This is because any user attempting to make use of the process will have to employ *P* and thereby violate the copyright held on *P*. But to allow this would make it possible to effectively *circumvent* the restriction imposed by § 102(b) and as a result, the arduous requirements of patent law. All of which is arguably in conflict with the public policy requirements of the patents and copyright clause of the U.S. Constitution.

Though not without its problems, the doctrinal concept of *merger* has been recruited by the courts to deal with this sort of circumvention of the restrictions imposed by § 102(b). The dispositive question in such merger determinations is whether there exists programming different from *P* that is capable, *as a practical and not purely theoretical matter*, of effectuating the process embodied in the coding.[18] If there is not, then *merger* applies and *P* is not copyrightable. Conversely, if there are such alternatives, then *P* is copyrightable because it is only one of many ways of effectuating the desired process. What's noteworthy here for current purposes is that the merger test for copyrightability makes essential reference to the process "embodied" in the computer program at issue even though that process is not, because of § 102(b), itself copyrightable. Thus, § 102(b) does not bar the *relevance* for copyrightability of "embodied" processes. Insofar as the Copyright Office may have thought otherwise in the context of synthetic biology, it was mistaken.

While the merger test applied to computer programming in the abstract appears straightforward, it can be frustratingly complex in actual applications—which leads us to the question of the application of merger in the context of synthetic biology.

MERGER, SYNTHETIC BIOLOGY, AND THE COMPLICATIONS THAT ARISE

As a starting point, it's evident that any application of *merger* in the case of synthetic biology *must* take into account the functional characteristics of the *combination* of DNA and intended chassis. This is because the merger test—conceived as analogous to its employment in the context of computer programming—will be whether *equivalent functionality* can be achieved by a DNA sequence other than that for which copyright protection is sought. In short, whether the DNA at issue is the only way to obtain the desired input-output augmentation *for the intended chassis*.

In terms of the idea-expression dichotomy, the *expression* is the DNA specification for which copyright is sought, and the corresponding *idea* is as follows:

> DNA sequences such that the incorporation of such sequences in the intended and specified chassis yields the desired augmentation of input-output capabilities.

It has been argued that, at least with respect to the Prancer DNA Sequence, "merger simply is not an issue" because of the redundancy of the genetic code.[19] In short, because of that redundancy there will be a corresponding set of equivalent codon *paraphrases* that presumably will lead to the same augmentation of input-output functionality as the original DNA sequence. But taking this tack while apparently promising ultimately leads to the unintended result that the copyright protection afforded to the DNA sequence would be of virtually no value since competitors could simply substitute one of the codon paraphrases as their own and thereby avoid the Prancer DNA copyright.

The assumption of equivalent effect (as in the case of the Prancer Sequence), however, does not hold in general because *codon optimization* is often required for maximum efficiency in the production of the intended output.[20] Therefore, in cases where optimization is required and there is a unique optimizing DNA sequence, the redundancy of the genetic code will not provide the alternative means required in order to show that there is no *merger* with the applicable *idea*. Accordingly, the *expression* merges with the *idea* and the DNA at issue is not copyrightable. This, too, is problematic, since it in effect penalizes developers of uniquely efficacious and optimized synthetic DNA. Intellectual property protection in such cases would have to be sought in the regime of patent law—not a welcome result for copyright enthusiasts.

Thus, where merger does not apply because of the availability of equally efficacious codon paraphrases, the copyright protection afforded will be marginal because competitors will be free to use such alternative codings. And where merger does apply, copyright protection is not available and intellectual property protection must be sought by way of patent. Thus, in either case copyright appears to offer little value. This unappealing result can be mitigated once the possibility of achieving equivalent functionality by means of DNA codings that are not paraphrases is taken into account. Here's how.

Begin by considering the case where there's a subset of the codon paraphrases, all of whose members are as functionally effective (when incorporated in the intended chassis) as the DNA initially specified. Here's the question: If the initial DNA is copyrightable, does that copyrightability transfer as a matter of law to all members of the subset—thus making the holder of an individual copyright also the holder of the copyright on all members of this subset of codon paraphrases?

In cases such as this, the concept of *literal* versus *nonliteral copies* can be profitably brought into play. The underlying motivation for the distinction is simple enough, namely, that mere *paraphrases* by a non-copyright holder constitute an infringement of the original copyright. The concept of a *nonliteral* copy is the legal generalization of the motivating notion of a paraphrase, while a *literal* copy is just what that term suggests, that is, a verbatim copying of the original expression. By contrast, a *nonliteral* copy is "paraphrased or loosely paraphrased rather than word for word" (*Lotus Dev. Corp.* 1995, 814). At a more general level, the notion is that nonliteral copies have in some sense *a shared content* that is sufficient to afford the nonliteral copy the same copyrightability as the original expression.[21] Thus, if a non-copyright holder employs a nonliteral copy of a copyrighted expression, that use of the nonliteral copy constitutes an infringement of the copyright.

In the case of synthetic biology, the applicability of the concept of a *nonliteral copy* is based on the relatively straightforward notion of a *paraphrase* involving the use of alternative codons for the same amino acid, and where the additional requirement for the necessary *shared content* is equivalent functional efficacy. So, if there's a subset of the codon *paraphrases* whose members are *as functionally effective* (when incorporated into the intended chassis) as the DNA initially specified, that subset constitutes the set of *nonliteral* copies, that is, *copies that warrant whatever copyright protection is afforded to the initial DNA specification*. That the copyright protection so extends to the nonliteral copies is a consequence of the fact that the initial DNA sequence and its nonliteral copies share all of the relevant features for copyrightability. Thus, if the initial DNA is, in fact, copyrightable, then the holder of the DNA sequence copyright ipso facto inherits the benefits of copyright protection over all the uses of the nonliteral copies.

Continuing along the same doctrinal vein, this set of nonliteral copies, that is, the equally efficacious codon paraphrases, corresponds to the following abstraction:

> DNA sequences *that are paraphrases of the initial DNA* such that incorporation of such sequences in the intended chassis yields the desired augmentation of input-output capabilities.

This abstraction thus constitutes an *intermediate idea* that is subsumed under the overarching and *merger* determining *idea*, specified earlier, namely:

> DNA sequences such that the incorporation of such sequences in the intended chassis yields the desired augmentation of input-output capabilities.

The merger test, therefore, will be whether the set of nonliteral copies is *exhaustive* of the means of instantiating this overarching idea. If so, then copyright protection must be denied for the original DNA at issue. See figure 2.4 for a graphical representation of the merger analysis and its interacting doctrinal components. This analysis of merger applied to the set of the nonliteral copies of a DNA sequence closely tracks that used by the court in *Oracle* (2014, 1354–56, 1367–68) regarding the merger of certain structural aspects of the computer programming at issue, but without the many twists and turns, and intervening complications of that decision.

The immediate advantage of this approach to merger is that it brings to the fore the relevance of a DNA sequence that is not a codon paraphrase but nevertheless yields equivalent functionality. If such a sequence exists, then the subset at issue of paraphrases with equivalent functionality does not merge with the merger-defining idea and the unwelcome consequences highlighted earlier may be avoided. Thus, copyright protection would be available for the Prancer DNA Sequence and its nonliteral copies if there exists a DNA sequence that is not a codon paraphrase that nevertheless yields equivalent functionality. Similarly, for situations where there is a subset (of one or more members) of optimized sequences.

But all is not clear sailing. In particular, the reader may wonder what determines whether a particular level of abstraction constitutes a merger determining idea versus a level of abstraction that does not. Why, for example, doesn't the intermediate idea as specified above serve as a merger-defining level of abstraction? *There is no clear-cut answer.* In *Apple*, the court noted that "[m]any of the courts which have sought to draw the line between an idea and expression have found difficulty in articulating where it falls," and explained that, when dealing with computer programming,

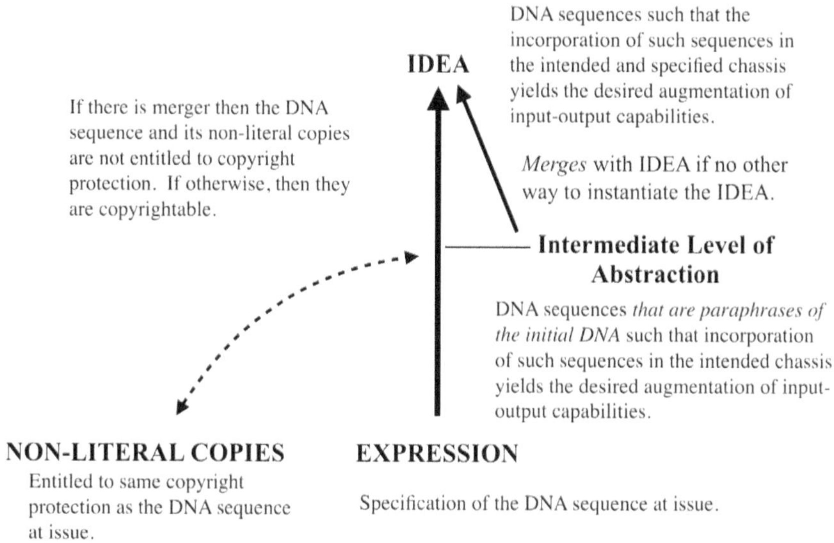

Figure 2.4 The Copyright Concepts of *Merger* and *Nonliteral Copy* and Their Application to the DNA Sequences of Synthetic Biology. *Source*: Author created.

"the line must be a pragmatic one, which also keeps in consideration the preservation of the balance between competition and protection reflected in the patent and copyright laws" (*Apple* 1983, 1253. Internal quotation omitted). But since this absence of a "bright-line" distinction is a general problem for copyright law, it should be expected also to be present when dealing with synthetic biology.

Problems more specific to synthetic biology, however, will arise in the application of the merger test. To demonstrate that the *expression* set of *nonliteral copies* does not merge with the overarching *idea*, one would have to show, as a practical matter, that a DNA sequence that was *not* a nonliteral copy could be engineered that would result in equivalent functionality when incorporated in the intended chassis. It is not apparent how this could be shown short of actually producing such a DNA sequence which would mean having to duplicate *mutatis mutandis* the original development program that resulted in the DNA sequence for which copyrightability was claimed in the first place. Whether less demanding means, based on theoretical analysis and analogies with similar cases, would suffice will depend—as in the case of patent applications—on the particularities of the case at hand. Conversely, to show that the relevant DNA expression merges with the overarching idea, one would have to demonstrate that theoretical and practical limitations make any alternative means of achieving equivalent functionality impossible. There are, however, two substantial scientific obstacles that stand in the way of such merger determinations.

First, the input-output functionality of the constructs of synthetic biology is, by and large, not digital but rather analogue in nature with a continuous range of responses depending on the specifics of the relevant initial conditions.[22] Consequently, there may not be clear-cut, digital criteria for otherwise equivalent satisfaction of the functional characteristics at issue in a merger determination. Insofar as optimization is at issue, for example, there may not be a unique measure of such optimization (such as in the case of being better in some respects but not others), and moreover, the optimization sought may be with regard to differing purposes.

Second, and as already noted above, the design of the constructs of synthetic biology is not fully predictive because, given the complexities and analog processes involved, the theoretical models employed in the design process are irredeemably idealized and consequently at best only approximate. Thus experimentation and subsequent fine-tuning are required to achieve the desired functionality. Consequently, the theoretical basis for making merger determinations will be compromised because of its idealized and approximate nature.

Even assuming that these obstacles could be overcome and made manageable in the context of a merger determination, having to deal with them will inevitably result in a *copyright regime that will have morphed into a close approximation of what already exists in patent law*. To see why, consider first the consequences for copyright registration.

Requirements for registration will have to be considerably more complex than just having to submit the specification for the DNA sequence at issue along with a showing of originality and a "modicum" of creativity. In particular, as shown above, because the DNA sequence by itself is just a non-copyrightable recipe for the construction of a useful article, and in order to avoid unnecessary post hoc litigation uncertainty and expense, (1) the performance and functional characteristics of the DNA-chassis composite should be *required for registration*, and (2) registration should be required *in all cases*.

Such registration requirements raise the question of whether a patent-like *justification* or *validation* of the claimed performance characteristics—especially those implicated in potential merger determinations—is to be required at the stage of copyright registration or postponed until needed in litigation. If the former, then copyright registration well and truly begins to merge into a close approximation of a patent application. If the latter, then we're left with a highly uncertain and potentially very expensive situation. In either case, there is at best only marginal simplicity and legal benefit to be gained by incorporating synthetic biology into copyright law.

There is yet another, and unavoidable, problem for registration, namely, that of the specification of the *range of applicability* of the sought-after copyright protection. Here the question will be whether the use of copyrighted DNA by a non-copyright holder would constitute an infringement if incorporated into a chassis *different* from what's specified in the original registration. In other words, whether the copyright applies only to the registered *combination* of DNA sequence and intended chassis or whether such copyright extends to the use of the DNA sequence *in all compatible chassis*. The urgency of the question is exacerbated by the fact that efficient chassis design is an ongoing and highly developed activity in its own right.[23] So as more compatible chassis become available, the question will be whether the copyright protection for an *already registered* DNA extends to its incorporation in newly discovered or developed compatible chassis.

This problem of determining the range of applicability does not arise with regard to computer source code because of another (though related) *disanalogy* between source code and the DNA specifications of synthetic biology, namely, that the functionality of programming source code is independent of its implementation in any particular machine—assuming, of course, effective implementation of the underlying language in the host computers.[24] This is why a specification of machine implementation—the computer programming analogue of chassis specification—is not required for the registration of computer source code.[25]

It would be highly inefficient if copyright were to be granted only for the specifically registered combination of DNA and chassis, that is, to require a separate copyright registration for the use of already copyrighted DNA in a different chassis. One way to deal with the undesirable redundancy of requiring different registrations for different chassis would be to require, *as in patent law*, an extensive review of the existing "art" and scientific underpinning, where that review would justify claims as to the claimed range of application of the DNA sequence at issue, that is, that of the claimed "invention," to use the terminology of patent law. But to follow along this path would push copyright even further into the realm of patent law.

SUMMARY AND CONCLUSION

The argument that the DNA sequences of synthetic biology should be considered as copyrightable is based on the claimed analogy between those sequences and computer programming, and the beguiling simplicity that would result, namely, that all that would be required for registration, and more generally copyrightability, would be a specification of the DNA sequence at issue along with a showing of originality and some "modicum" of creativity. The claimed analogy, however, must be tempered because of a significant *disanalogy* that, unlike computer source code, the emergent functionality of a DNA sequence—like that of ENIAC's hard-wired programming—is invisible on its face. Consequently, reference to the composite of DNA and its functional consequences once incorporated in its chassis is necessary for copyrightability. Otherwise, the DNA sequence is not copyrightable because, by itself, it is no more than a recipe for the construction of at best a "useful object" to be incorporated in an unspecified biological chassis yielding unspecified functional consequences. Once the necessity of the composite system of DNA and chassis is brought into play, and the "embodied" functional characteristics thereby made visible, the doctrinal concepts of copyright law, including that of merger, can be shown to operate in analogous fashion—at least in the abstract—to their employment with regard to computer programming. Delving more deeply, however, into the practical and scientific problems raised in the application of these concepts in the case of synthetic biology reveals an unavoidable slippery slope of complications that eventually leads to and demands a copyright system that closely approaches that of the patent system. The claimed benefits of simplicity and economy to be obtained by making copyright applicable to synthetic biology are thus fatally compromised. Accordingly, there is no reason based on sound public policy for so bloating the existing legal system. Intellectual property protection for synthetic biology should remain in the domain of patent law.

It should be clear that the question of whether copyright law should be made applicable to synthetic biology is exceedingly complex with many interwoven strands of law, technology, and commercial interests, all subject to Constitutional constraints as well the requirements of sound public policy. Getting the right answer is a task where many hands have been and will be involved—and I have tried to indicate how and why this is so. What professor Pitt adds to this complex mix, and others like it involving "mankind's interaction with the world," is his assertion that philosophers have an *obligation* to contribute to the solution of such problems and not rest content with issues that are parochial and narrowly professional. Acting on such an obligation in the ordinary course of academic life will be difficult but not impossible. As an aspirational goal, it is to be taken seriously.

NOTES

1. See, for example, Rai and Boyle (2007), Ledford (2013), Nelson (2014), and Holman (2017, 405–12, 420, 455–56).

2. For an example of a synthetic biology patent at the more accessible end of the complexity spectrum, see U.S. Patent No. 9,365,603 B2 (issued June 14, 2016).

3. For a brief review of how this happened, see *Apple Computer* (1983, 1247–49).

4. For the details on the application and its rejection by the U.S. Copyright Office, see Holman, Gustafsson, and Torrance (2016).

5. For a review of the powers of and deference due the Copyright Office as an administrative agency, see Burk (2018).

6. For a comprehensive introduction, see Baldwin (2012).

7. For the details on the modular interaction of the *lac* operon with its chassis, see the exquisitely well done Ptashne and Gann (2002).

8. See Goldstein and Goldstein (1982) and Burks and Burks (1981).

9. See, for example, Holman (2017): "The information content encoded by a synthetic genetic sequence . . . is *entirely analogous* to representations of computer code that are directed toward a human reader rather than a machine, as printed *source code* might be read from paper" (Holman 2017, 402, emphasis added).

10. U.S. Copyright Office (2017, § 607) and U.S. Copyright Office (2018) stating that "[s]uch [object code] registrations are made under the Office's rule of doubt since the Office cannot determine with certainty the presence of copyrightable authorship" (U.S. Copyright Office 2018, 2).

11. For more details on the bias in copyright law against utilitarian objects and the artistic feature exception, see *Star Athletica* (2017, 1007, 1016) holding that "[a] feature incorporated into the design of a useful article is eligible for copyright protection only if the feature (1) can be perceived as a two- or three-dimensional work of art separate from the useful article, and (2) would qualify as a protectable pictorial, graphic, or sculptural work—either on its own or fixed in some other tangible medium of expression—if it were imagined separately from the useful article into which it is incorporated."

12. Figures 2.2 and 2.3 (adapted) are with respect to Part BBa_K389050 from The Registry of Standard Biological Parts.

13. The genetic "parts" as notated in figure 2.3 capture standardized, general operational features and thus each such part may be actualized by differing sequences of DNA. See Baldwin et al. (2012, 73–88) for more details. The specific correspondence, for example, between the DNA sequence and its "embodied" functional parts for Part:BBa_K389050 is available at The Registry of Standard Biological Parts.

14. See Baldwin et al. (2012, 89–107), Cardinale and Arkin (2012), Knuuttila (2013), and Li and Borodina (2015, 9–10).

15. For a concise review of the use of analogy by Congress and the courts, see Holman (2011, 711–12) and Holman (2016, 106).

16. While Holman (2017) reviews many of the desirable principles to be included in a sui generis version of copyright tailored for synthetic biology, he does not recognize the consequential disanalogy between specifications of synthetic DNA and

computer source code, what I have described as the invisibility of the "embodied" functional properties of DNA when viewed in isolation from its intended chassis. Consequently the many complications caused by this invisibility are not taken into account.

17. Which is how (whether fairly or not) the Court of Appeals for the Federal Circuit understood the district court's opinion in the *Oracle v. Google* litigation. See *Oracle America* (2014, 1366–67).

18. In addition to "being feasible to within real-world constraints" (*Lexmark International* 2004, 536), such practical possibilities "are to be evaluated at the time of creation, not at the time of infringement" (*Oracle America* 2014, 1361).

19. See Holman, Gustafsson, and Torrance (2016, 116). Making reference to merger, as Holman et al. do here, requires that specific reference be made to the functionality of the *composite* system of DNA and the intended chassis. Hence, the copyrightability of the DNA requires more than just reference to the DNA sequence itself—which is exactly my point.

20. For a brief account and examples, see Baldwin et al. (2012, 45–46, 113–14). For more comprehensive reviews, see Webster, Teh, and Ma (2017) and Mauro and Chappell (2014). I should also note that Holman in a later paper (Holman 2017, 442) recognizes that optimization may disqualify a DNA sequence from copyrightability because of merger.

21. See *Lotus* (1995, 814) and *Oracle* (2014, 1557–58) for a discussion of the "abstraction, filtration, and comparison" test to determine the nature of the shared content in cases dealing with computer programming.

22. For a simple example of such analogue response data, see, for example, The Registry of Standard Biological Parts, Part:BBa_F2620, available at: http://parts.igem.org/wiki/images/f/fe/EndyFig1-F2620DataSheet.pdf. For an example of such analogue performance data in the context of a patent application, see the patent referenced above at note 2.

23. See, for example, Jansen et al. (2017), Redden, Morsem, and Alper (2015), Steensels et al. (2014), and Walker and Pretorius (2018) for reviews of the extensive research and development that have been devoted to more efficient variants of yeast to serve as a chassis.

24. Because of this independence, the concept of a *nonliteral element* of a computer program naturally evolved in the law. See *Oracle* (2014, 1354–56). The concept (which is to be distinguished from that of a *nonliteral copy*) arose in order to deal with questions of the copyrightability of certain *structural features* (what had been referred to as the "structure, sequence, and organization") of the programming as opposed to the programming code itself. This is because those features can be abstracted from the programming *without reference to the implementing host computer*. The question that arises then is whether these nonliteral (structural) elements can be copyright protected in a way that avoids the prohibition of § 102(b). The many difficulties with such an approach came to a head in the highly contentious *Oracle v. Google* litigation. The analogue, such as it is, in synthetic biology is with the *modularity* of the functional features of the DNA (as reflected in its "genetic parts") once incorporated in its intended chassis. See, for example, figure 2.3 which displays the genetic parts. Such parts and resultant modular functionality, of course, are not independent of

chassis selection. But to argue for copyrightability at this level takes us far beyond the simplicity of the original proposal for copyrightable DNA and even further into the realm of a patent-like legal regime.

25. See U.S. Copyright Office (2017, §§ 721, 1509(C)).

REFERENCES

Cases

Apple Computer, Inc. v. Franklin Computer Corp., 714 F. 2d 1240 (3d Cir. 1983).
Association for Molecular Pathology v. Myriad Genetics, Inc., 569 U.S. 576 (2013).
Baker v. Selden 101 U.S. 99 (1879).
Feist Publications, Inc. v. Rural Telephone Services Co., Inc., 499 U.S. 340 (1991).
Lexmark Int'l, Inc. v. Static Control Components, Inc., 387 F.3d 522 (6th Cir. 2004).
Lotus Dev. Corp. v. Borland Int'l, 49 F.3d 80 (1st Cir. 1995).
Oracle America, Inc. v. Google Inc., 750 F.3d 1339 (Fed. Cir. 2014), cert. denied, 135 S.Ct. 2887 (U.S. June 29, 2015) (No. 14-410).
Star Athletica, LLC v. Varsity Brands, Inc., 580 U.S. ___ (2017), 137 S. Ct. 1002.

Constitutional Authority, Statutes, and Other Government Documents

H.R. Rep. No. 94-1476 (1976).
U.S. Copyright Act of 1976, 17 U.S.C. (as amended 2016).
U.S. Copyright Office, Compendium of Practices (2017).
U.S. Copyright Office, Help: Deposit Copy (2018). Available at https://www.copyright.gov/eco/help-deposit.html.
U.S. Const. Art. I, Sec. 8, Cl. 8.
U.S. Patent Act, 35 U.S.C. §§ 1-376 (1994).
U.S. Patent No. 9,365,603 B2 (issued June 14, 2016).

Registry of Standard Biological Parts

The Registry of Standard Biological Parts, Part:BBa_K389050, available at http://parts.igem.org/Part:BBa_K389050.
The Registry of Standard Biological Parts, Part:BBa_F2620, available at http://parts.igem.org/wiki/images/f/fe/EndyFig1-F2620DataSheet.pdf.

Articles and Books

Baldwin, G. et al. 2012. *Synthetic Biology—A Primer*. London: Imperial College Press.
Burk, D. L. 2018. "DNA Copyright in the Administrative State." *UC Davis Law Review* 51: 1297–349.

Burks, A. W., and Burks, A. R. 1981. "The ENIAC: First General Purpose Electronic Computer." *Annals of the History of Computing* 3: 310–99.

Cardinale, S., and Arkin, A. P. 2012. "Contextualizing Context for Synthetic Biology—Identifying Causes of Failure of Synthetic Biological Systems." *Biotechnology Journal* 7: 856–66.

Goldstine, H. D., and Goldstine, A. 1982. "The Electronic Numerical Integrator and Computer (ENIAC)." In *The Origins of Digital Computers* (3rd ed.), edited by Reprinted in B. Randell, 359–73. Berlin: Springer Verlag.

Holman, C. 2011. "Copyright for Engineered DNA: An Idea Whose Time Has Come?" *West Virginia Law Review* 113: 699–738.

Holman, C. 2017. "Charting the Contours of a Copyright Regime Optimized for Engineered Genetic Code." *Oklahoma Law Review* 69 (3): 399–456.

Holman, C., Gustafsson, C., and Torrance, A. W. 2016. "Are Engineered Genetic Sequences Copyrightable?: The U.S. Copyright Office Addresses a Matter of First Impression," with attached "Supplementary Document 1: Request for Reconsideration of Denial of Copyright Registration of Prancer DNA Sequence," and "Supplementary Document 2: Affirmance of Refusal for Registration." *Biotechnology Law Report* 35 (3): 103–24.

Jansen, M. L. A. et al. 2017. "*Saccharomyces cerevisiae* Strains for Second-Generation Ethanol Production: From Academic Exploration to Industrial Implementation." *FEMS Yeast Research* 17 (forthcoming). doi: 10.1093/femssyr/fox044.

Knuuttila, T. 2013. "Basic Science through Engineering? Synthetic Modeling and the Idea of Biology-Inspired Engineering." *Studies in History and Philosophy of Biological and Biomedical Sciences* 44: 158–69.

Ledford, H. 2013. "Bioengineers Look beyond Patents: Synthetic-Biology Company Pushes Open-Source Models." *Nature* 499: 16.

Li, M., and Borodina, I. 2015. "Application of Synthetic Biology for Production of Chemicals in Yeast *Saccharomyces cerevisiae*." *FEMS Yeast Research* 15: 1–12.

Mauro, V. P., and Chappell, S. A. 2014. "A Critical Analysis of Codon Optimization in Human Therapeutics." *Trends in Molecular Medicine* 20 (11): 604–13.

Nelson, B. 2014. "Synthetic Biology: Cultural Divide." *Nature* 509: 152–54.

Pitt, J. 2016. "The Future of Philosophy: A Manifesto." In *Philosophy of Technology after the Empirical Turn* (Philosophy of Engineering and Technology; Vol. 23), edited by M. Fransen, P. E. Vermaas, P. A. Kroes, and A. W. M. Meijers, 83–92. Switzerland: Springer.

Ptashne, M., and Gann, A. 2002. *Genes & Signals*. Cold Spring Harbor: Cold Spring Harbor Laboratory Press.

Rai, A., and Boyle J. 2007. "Synthetic Biology: Caught between Property Rights, the Public Domain, and the Commons." *PLoS Biology* 5 (3): e58. https://doi.org/10.1371/journal.pbio.0050058.

Redden, H., Morsem N., and Alper, H. S. 2015. "The Synthetic Biology Toolbox for Tuning Gene Expression in Yeast." *FEMS Yeast Research* 15: 1–10.

Steensels, J. et al. 2014. "Improving Industrial Yeast Strains: Exploiting Natural and Artificial Diversity." *FEMS Microbiology Reviews* 38: 947–95.

Walker, R. S. K., and Pretorius, I. S. 2018. "Applications of Yeast Synthetic Biology Geared towards the Production of Biopharmaceuticals." *Genes* 9: 340–62.

Webster, G. R., Teh, A. Y., and Ma, J. K. 2017. "Synthetic Gene Design-The Rationale for Codon Optimization and Implications for Molecular Pharming in Plants." *Biotechnology and Bioengineering* 114 (3): 492–502.

Chapter 3

A Defense of Sicilian Realism

Andrew Wells Garnar

This chapter is about the varieties of scientific realism and thinking through relationships between masters and disciples. Examining the question of whether the scientifically real should be understood reductively or expansively provides an opportunity to explore what debts Joseph Pitt owes Wilfrid Sellars, and, in turn, my own inheritance received from both. Although Joseph Pitt never formally studied under Wilfrid Sellars, there remains a sense in which Pitt is something of a student of Sellars. Pitt's dissertation was on Sellars, and he remains a recurring figure in many of Pitt's publications over the years. From personal experience, I can attest to the deep impact that Sellars made on Pitt. In ways similar, Pitt has left his impact on me. Here we have master and disciple, further complicated by the disciple becoming a master in his own right to his own disciples.

To be a disciple requires putting oneself in a position of subservience to the master such that the disciple can internalize the lessons of the master. Yet, as Jacques Derrida reflects with respect to his relationship with Michel Foucault, this produces an "unhappy consciousness" within the disciple.

> Starting to enter dialogue in the world, that is, starting to answer back, he always feels . . . like the "infant" who . . . cannot speak and above all must not answer back. And when, as is the case here, the dialogue is in danger of being taken—incorrectly—as a challenge, the disciple knows that he alone finds himself challenged by his master's voice within him that precedes his own. He feels himself indefinitely challenged, or rejected or accused; as a disciple, he is challenged by the master who speaks within him and before him, to reproach him for making this challenge and to reject it in advance, having elaborate[d] it before him; and having interiorized the master, he is also challenged by the disciple that he himself is. This interminable unhappiness of the disciple perhaps stems from

> the fact that he does not know—or is still concealing from himself—that the master . . . may always be absent. The disciple must break the glass, or better the mirror, the reflection, his infinite speculation on the master. And start to speak. (Derrida 1978, 31–32)

If the disciple is ever to enter into a genuine dialogue with their master, they can no longer merely repeat the words of the master, let the master's internalized voice speak through the disciple's mouth. Dialogue requires different voices speaking. For the disciple to engage in such a dialogue, always risking that it is seen as a challenge, requires that the disciple find their own voice "and start to speak." The difficulty here is twofold: it means *both* breaking with the master *and* the master's voice the disciple has internalized. When the disciple does this, is it a failure? If so, is it a failure on the part of the student? The master? Both? If there is no failure per se, then what did the master actually teach? In what sense, other than in terms of recounting pedigrees, is the disciple actually a disciple and the master actually a master? At least with Pitt, I can attest that if the student breaks with him *and* can provide reasons for this, he sees this not as a failure but as a success (having never had the opportunity to meet Sellars, I cannot say whether he would agree with Pitt on this). For him it is a significant success, because this means that Pitt provided an education that allowed the disciple to surpass him. The disciple learned enough from him, not in the sense of this or that doctrine, but in terms of how to do philosophy, to move beyond what was taught.

With these musings on the relationship between master and disciple in mind, let me now turn to the heart of this chapter: scientific realism. This is one place where Pitt and Sellars have what turns out to be more than a minor disagreement. In fact, by pulling at this disagreement, one might be led to wonder whether they live in the same conceptual universe. "Sicilian realism" is Pitt's name for his particular species of scientific realism. It is at once very important for understanding the overall trajectory of his thought, a fascinating idea, and one he would agree is among his less developed. Summarized quickly, it is a form of realism that holds that all the entities referred to in mature scientific theories are real. This is opposed to the more traditional scientific realism of Sellars which holds that the most fundamental entities of physics (FEPs) are real.[1] Both of these are forms of scientific realism, holding that science provides an account of the actual structure of the world, not mere calculation devices or social constructions. But each provides a very different understanding of what "realism" means, hence a different understanding of what the world is actually like. Traditional realism is like an austere desert, holding to bear the minimum of entities; the other is like a jungle teeming with life. Sellars defended the traditional sort and went so far as to anticipate a move like Sicilian realism, and attempts to cut this off in favor of a more reductionist account. While Pitt offers brief sketches of Sicilian realism in his writing, his most substantive discussion appears in the final pages of his *Thinking about Technology*, and what he offers there is little more than a sketch.[2] It is a rather provocative sketch for those who find something puzzling with more traditional reductionist forms of scientific realism, but lacks a systematic articulation.

Given Pitt's debt to Sellars, we find a variation on the issue discussed above. How faithful of a disciple of Sellars is Pitt if he breaks with Sellars on this crucial point of scientific realism? The challenge is further compounded because, as I will show, the reason Pitt and Sellars break on the nature of scientific realism is because of other, deeper philosophical differences between them. Can we think of Pitt as a disciple of Sellars at all? I shall briefly return to this, and my own relationship to Pitt's work, in the conclusion of this chapter. My immediate concern is offering something of a defense of Sicilian realism by working through the criticisms that Sellars raises against this sort of position. To this end, I will briefly summarize the core claims of Sicilian realism and then lay out the Sellarsian case against Pitt's sort of realism. From here I begin to explore how Pitt responds to traditional scientific realists, looking first at where he thinks they go wrong and why taking technology and practice is decisive for correcting such mistakes. The shift to understanding science as shot through with technology ends up being the cornerstone of defending Sicilian realism.

A SIMPLIFIED VERSION OF A COMPLICATED IDEA

Sicilian realism is a thoroughly anti-reductionist species of scientific realism. According to Pitt,

> "Common to all versions of scientific realism is the notion of reductionism. That is, whatever else is the case, the bountiful variety of *things* and/or *processes* is to be understood as deriving from the fundamental things or laws that the realism endorses" (Pitt 2000, 134–35. Emphasis original).

Instead of this reductionist move, equating what is real and what is fundamental, Pitt breaks these apart. Questions of what is real hang separately from those of what is fundamental. Those entities which are postulated by well-supported scientific theories are real. All of them.

Instead of the minimalist ontology of most philosophers of science, including Wilfrid Sellars, Pitt's world can be described in the words of William James "as one great blooming, buzzing confusion" (James 1950, 488). What deserves the honorific of "real" is not some very basic set of entities proposed by physical theory. Or at least not *just* these entities. The levels of composite entities made from these fundamentals also deserve to be called "real" as well. In which case, the world echoes another of James's pithy statements: "profusion, not economy, may after all be reality's key-note" (James 1987, 570). One of Pitt's reasons for referring to this as Sicilian realism underlines this complexity. The name reflects

> "the rich and varied cultural history of Sicily. For centuries, Sicily was at the center of the trade routes and invasion paths. . . . The composition of Sicilian culture reflects Greek, Roman, German, and Moorish input, among others. In its architecture, music, food, and customs it is a multilayered and complex place, much like our universe" (Pitt 2000, 135n11).

One could also find illustrative contrasts in art. Instead of the minimal worlds of Ellsworth Kelly or, better still, Mark Rothko (because the latter reliably used more than one color), Sicilian realism understands the world as something by Jean Miro, in particular a painting like *The Harlequin's Carnival* with its wild variety of figures contained on one canvas, or perhaps a Hieronymus Bosch triptych. These give a sense of the image of the world that Pitt relies upon. It is anything but simple. It is a world that contains both the macroworld humans tend to take for granted, the super-macroscopic (solar systems, galaxies, etc.), and the microscopic (nerves, cells, genetic material, molecules, etc.), a rather confusing place when taken all together.

A Sicilian realist holds that the entities of mature scientific theories are real. Where Pitt breaks from traditional realists is with the scope of those entities regarded as real. A realist like Sellars holds that the real is only unexplained explainers, those entities that everything else should be reducible to. In which case, it is not simply quarks and leptons that are real, or strings or whatever defines the FEPs of the universe. According to Pitt, quarks and leptons are real (the jury is out still about strings). But so are atoms, molecules, macromolecules, cells, bodies, continents, stars, and galaxies. This is the crucial point: Sicilian realism proposes that scientific entities are real all the way down. In this way, it is "realism with a vengeance" (Pitt 2000, 135), thus providing another reason why Pitt uses this particular name. Central to Pitt's realism is that the world is a very complicated place that can be "cut" at any number of different joints. Different sciences approach the world at different levels, revealing different entities of which it is comprised.[3] Rather than the traditional realist move of consistently relying on reducing a higher level to lower levels, Pitt suggests that the entities at each of these levels are real.

Part of the appeal of Sicilian realism, to my mind, is the extent to which it allows a scientific realist to have their cake and eat it too. Pitt's scientific realism allows one to take seriously (i.e., treat as real) both the entities of science and the world of common sense. My saying this should not be taken as an endorsement of the world of common sense simpliciter (or something like dualism either). There are grounds for being critical of important parts of this world, both for reasons coming out of science and those created by common sense itself. *The concern here is not to preserve the world of common sense altogether, but rather to resist a move made by some defenders of scientific realism to jettison the world of common sense in favor of the world of science.* Sellars refers to these as the "manifest image" and "scientific image," respectively, and makes precisely this move with respect to questions of realism (the manifest image is important for other reasons, but not in terms of describing the contents of the world). At a practical level, Sellars admits that it is ultimately unnecessary to attempt a wholesale replacement of the objects of common sense (like objects existing in Space and Time) (Sellars 1991, 97). Admitting that to seek to systematically replace the physical objects of the manifest image with those of the scientific image "because the better is the enemy of the best" (Sellars 1991, 97), Sellars commits himself to such a replacement ideally, that is to say in principle. Accepting even this self-imposed limitation, I will propose that there remains something important holding to the reality of the framework of common sense. Part of the motivation for this is to ensure that one's anti-foundationalism is as thorough as possible. One concern, which I only mention here, is that by taking the world of science as the only reality, this takes the scientifically real as something foundational, a given (thus falling prey to Sellars's own argument in *Empiricism and the Philosophy of Mind*).[4]

THE SELLARSIAN CHALLENGE

In an attempt to better understand and articulate a defense of Sicilian realism, I will now detail the challenge Sicilian realism faces from Wilfrid Sellars's very reductionist form of scientific realism. What might be Sellars most poignant statement comes from *Empiricism and the Philosophy of Mind*:

> But, *speaking as a philosopher*, I am quite prepared to say that the common sense world of physical objects in Space and Time is unreal—that is, there are no such things. Or, to put it less paradoxically, that in the dimension of describing and explaining the world, science is the measure of all things, of what is that it is, and of what is not that it is not. (Sellars 1991, 173. Emphasis original)

In this passage, Sellars eliminates the common-sense world of trees, squirrels, and human bodies (but not persons) in favor of the image that science provides. In several later essays, he clarifies what this position entails. For example, in "Phenomenalism," he proposes that the entities of science function like Kant's noumena, with the rather significant difference that these "things-in-themselves" are not unknowable. Rather, they are accessible only through scientific inquiry. The analogy he draws to Kant's noumena gives a sense of how unreal the common-sense world is:

> "the perceptual world is phenomenal in something like the Kantian sense, the key difference being that the *real* or 'noumenal' world which supports the 'the world of appearances' is . . . the world as construed by scientific theory" (Sellars 1991, 97, emphasis added).

Reality is found beyond the phenomenal world.

Sellars then further refines this stance in "Philosophy and the Scientific Image of Man." It is the developments here that concern me most. It is here that Sellars provides the most sustained discussion of the basics of the scientific image and its relationship to the world of common sense. He does not back off from the earlier reductionism but instead qualifies it. In his explication of the scientific image, he distinguishes two senses in which objects are composed of scientific entities. The first is that each domain described by science (such as atomic physics, chemistry, biochemistry, neuroscience) is wholly reducible to the final theory of physics. This clearly will not do. What goes on at the level of a biological cell cannot be adequately accounted for by quantum mechanics. For this reason, he endorses the view that while a cell or the brain or a molecule really is nothing other than atoms (or whatever FEPs there turn out to be), these higher-level objects do possess properties distinct from their constituents in lieu of their arrangement. This line of reasoning applies to the unobservable entities postulated by scientific theories. Where Sellars draws a clear line is between these scientific entities and those of common sense, what he refers to here as the "manifest image." These manifest objects, the "objects of Space and Time" discussed previously, are absent from the scientific image. Sellars excludes these objects of the manifest image because they possess properties that are fundamentally at odds with those of the scientific image.

It is in this context that Sellars worries about a position akin to Sicilian realism. Given the clash between the manifest and scientific images, he envisages three possible responses. The one he endorses is eliminating the manifest image as a *description of the world* in favor of the scientific image. Another he does not explore here is giving primacy to the manifest image and to take the scientific image as a "calculation device" to ease navigation of the manifest (see the aforementioned "Phenomenalism" for his sustained criticism of this sort of instrumentalism). The remaining option is to take the contents of both images as real. Sellars spends a bit of space showing why this option, although reasonable at first pass, falls apart on further inspection. The difficulty revolves around the differences between the objects of the images. The objects, including at the most fundamental level of the scientific image, are particulate.

Sellars clearly worked with a mid-twentieth-century conception of atoms as composed of discrete particles like protons, neutrons, and electrons. That noted, in the almost sixty years since the publication of "Scientific Image," the FEPs might be stranger, FEPs remain particles nonetheless, in sharp contrast to the objects of a manifest image that can be homogenous and/or continuous in ways that particles cannot. The quintessential example of this is Sellars's pink ice cube. When looking at the ice cube, although the cube itself is discreet and bounded, the pink color is distributed continuously throughout the cube. The pinkness cannot be accounted for in terms of discreet parts. Its fundamental homogeneity and the particulate nature of the entities that comprise the pink ice cube cannot be wedded together in any obvious way. For this reason, we see why Sellars likened the fundamental entities of the scientific image to Kant's noumena. They explain why it is the objects appear to humans, although they bear no resemblance to those objects. The pink ice cube, as a human sees it, is simply appearance, a Kantian phenomenon, produced through the interaction of those entities with the human perceptual apparatus. The pinkness is merely something that appears to humans but is not actually real. Thus, at a minimum, the contents of the world of "common sense" cannot be regarded as real. And given Sellars's commitment to a particular sort of reductionism, it is clear why Sicilian realism is a dead letter.

THE MYTH OF SIMPLICITY

In the face of Sellars's refutation, Pitt has several options. The first is to walk back the scope of Sicilian realism: to say that Sicilian realism applies only to the entities of Sellars's scientific image and then fight it out with Sellars over whether reductionism is warranted. This option is open to Pitt, yet sits uneasily with me, and I suspect did with him as well. I will mention two reasons. First, Pitt makes passing reference to Eddington's table. The question posed by Arthur Eddington was, which table is real: the one he sees or the one described by atomic physics? With reference to this choice, he states: "I want to argue that we don't eliminate the table in favor of a set of electrons" (Pitt 2000, 135 36). If taken at face value, this statement implies that Pitt does in fact argue that the world of experience is as real as electrons (and everything else contained in the scientific image). Furthermore, a criticism that Pitt makes of many scientific realists is their tendency to fetishize unobservable scientific entities at the expense of *observable* scientific entities. For example, one distinctive feature of Sellars's scientific image is the postulation of micro-entities that explain the behavior of the manifest image. This ignores important classes of macro- or super-macroscale scientific entities. Putting aside the question of observability for the moment, not all scientific entities are microscopic. For example, consider tectonic plates, stars, galaxies, weather systems, and ecologies. Some of these might not be observable in a standard sense, but none of these are microscopic like cells, molecules, or electrons. Given the caveats mentioned earlier, Pitt clearly takes these as real (especially those revealed by telescopes. See Pitt 2011, 90–93). This creates at least some dissonance between Pitt's position and that of Sellars.

Since Pitt is unlikely to capitulate to Sellars on this point, he needs a further argument. His response comes in the form of what he calls "the Myth of Simplicity" (Pitt 2010). This myth involves running together what is "fundamental" with what is "real," because those "fundamental" things are supposed to be the "simplest" (in some sense or other). He has no issue with the idea that scientific theories describe real phenomena, regardless of whether they are observable or unobservable. He rejects *the claim that only that which is most fundamental is real*. How does he make this argument?

Pitt's critique of the Myth of Simplicity involves two moves. The first is the assertion that the appeal of simplicity is simply a dogma among philosophers of science. The other is the replacement with this dogma with an appeal to complexity. With the first, he admits that while the equations that explain the beginning of the universe *might* be simple, this in no way requires that everything else in the universe is, in fact, reducible to this simplicity. Two comments about this point. First is that the simplicity that these equations might possess must be understood in context. By this I mean that, within the context of certain sorts of physics, if such equations exist, they count as simple in comparison with other options (real or hypothetical). It is not clear what "simple" in the abstract might mean. This is compounded by the relative complexity of many physics equations for the uninitiated. For those who understand them, they might be simple, but this is not obvious to those outside of the field. Second, even if we put aside the question of what defines "simplicity," the more substantive claim from Pitt is that, even if there are such "simple" equations, "it doesn't follow that the universe as we know it now must abide by principles of simplicity" (Pitt 2000, 135). Why? Again, Pitt only hints at what the argument is here, and what is here seems open to criticism.

What Pitt is *not* arguing here is that larger entities are not composed of smaller entities, and ultimately FEPs. This would be a rather odd thing for a scientific realist to claim. Pitt agrees that, inasmuch as the standard model is well supported and fits with other theories, quarks and leptons are the fundamental building blocks of the world, and that perhaps these FEPs will be replaced by strings. This issue is not that of what is most fundamental. Instead, it is the conflation of "most fundamental" with "most real." He finds this conflation, which is a form of reductionism, to be endemic to all versions of scientific realism. And that is indefensible, noting that he does "not believe that there is a justification for the methodological appeal to reduction" (Pitt 2000, 135). Part of what Pitt claims here is that just because the fundamentals are simple does not entail higher-level entities are simple as well, and seems uncontroversial. This insight seems to be the general idea of complexity theory (Mitchell 2009), and there is evidence that Sellars does not disagree with it (see Sellars and Meehl 1956). What is more problematic is the shift from this point about simplicity to the claim that simplicity entails reductionism. Without digging deeper, this appears to be a mere assumption.

A SELLARSIAN REJOINDER

At this stage, I expect that Sellars would be unmoved and should pose two rejoinders to Pitt. The first is that, despite all these claims to reality possessing many levels, there is nothing here that solves the problem he raised about linking the microworld of theoretical science to the world of common sense. On this point I readily admit that this is a fair criticism. There is still something of a mystery about how the particulate world of the scientific image produces the phenomenological experience of pink ice cubes and everything. The best I can offer here is a hope that one day this problem might be resolved.[5]

The second is challenging in a rather different way. The rejoinder is, "Simply labeling something a myth does not make it mythical." What Pitt appears to argue is that when scientists work with composite objects (like molecules, genes, cells, etc.), they approach these objects as if they are real. Because they act as if these composite objects are real, this gives the appearance that the universe is composed of many different levels that are all real. But simply because scientists working at levels above that of the FEPs do not conceive of the objects of their inquiry in terms of the FEPs does not mean that their objects of inquiry are, at base, nothing more than these FEPs. In practice, Pitt is correct that scientists *take* this or that as real, but that does not entail that they are *really real*." This criticism goes to the heart of Pitt's Sicilian realism, since it denies taking these levels other than those FEPs as *really real*.

THEORY, PRACTICE, AND REALISM

The response I will make to Sellars and in line with Pitt exposes a deeper fissure between the two philosophers. I began by tracing out the differences between them about the role of reductionism in scientific realism. What produces this divergence involves the significance of "theory." Sellars's philosophy falls squarely within what Ian Hacking (1983) referred to as "representation." Representational philosophies of science begin and end with the analysis of scientific *theories*. Some philosophers of science dealt with questions of how evidence (understood in a rather as antiseptic sense at their use of "theory") verified, falsified, or corroborated theories. Sellars did some work on this, but much of his writing was devoted to issues about language, reference, perception, intentionality, and the relationship between scientific and other languages. Still working at the level of representations, or a "mummified" image of science to use Hacking's more acidic terminology, but not quite what other philosophers of science were up to (Hacking 1983, 1–4). Pitt's work is closer to what Hacking referred to as "interventional" philosophy of science, inasmuch as Pitt is concerned with doing things with instruments and in experiments. This difference can be explained, in part, as generational. For philosophers of science in Sellars's generation, the actual practice of science was hardly of central concern. Pitt comes of age in the following generation. Pitt finished his bachelor's in 1966—shortly after the 1963 publication of Thomas Kuhn's *Structure of Scientific Revolutions*—and his doctorate in 1972, in the wake of Imre Lakatos and Alan Musgrave's edited volume *Criticism and the Growth of Knowledge* (1970). This, to hear Pitt tell the story, is what really made a segment within the philosophy of science community take note of Kuhn and consider the implications of *Structure*. The time during which Pitt received his degrees marked a change in the sensibility in the philosophy of science, ushering in a different image of science, one that is much less pure and much more messy. Since that time, Pitt has not simply been caught up in this shift, but helped to further move the philosophy of science in this direction (as this volume helps demonstrate) and to move the philosophy of technology toward instrumentation.

This generational change helps make sense of some of the differences in philosophical commitments but does not go far enough. The shift from focusing on theory to taking practice more seriously is not simply another example of changing academic fashions. Pitt, along with other philosophers of science like Allan Franklin, Peter Galison, Ian Hacking, and Helen Longino, demonstrated how emphasizing the practice of science shifted one's understanding of the very basics of science itself (Pitt 1995). Rather than the more theory-driven understanding that dominated the philosophy of science for most of the twentieth century, Pitt and these other "new experimentalists" emphasize the significance of practice within science, in particular, as their moniker would lead you to expect, experiments. This shifts considerations of what is real. No longer is the principal concern about what counts as real determined by theoretical commitments, but rather what can be manipulated in the context of scientific practices. At least without a good bit more argument, this move to emphasizing practices does not eliminate the older theory-driven approach to the philosophy of science, showing its ultimate futility. Rather, it diminishes the importance of theory in thinking about science, showing that the appeal to theory is not the only game in town for determining what is scientifically real.

This digression about the shift from theory to practice points the way toward meeting Sellars's second criticism of Sicilian realism; it does not address what is *really* real. Pitt's concern with practice shifts questions about realism from ones about the relationship of sentences to other sentences, and hopefully, at some point to the actual world, to ones about what scientists do in the laboratory and the field. This concern for practice re-centers the issue of realism into a pragmatic space of action, doing things. "Pragmatic" here is used both in the everyday sense of "practical," and with reference to the philosophical movement. The fingerprints of American pragmatism are found all over Pitt's work. He makes frequent reference to C. S. Peirce, though one can find the spirits of William James and John Dewey at work as well (never mind latter-day, more analytically oriented pragmatists like C. I. Lewis, Nelson Goodman, and Wilfrid Sellars himself). Because Pitt is not only a disciple of Sellars, but pragmatism as well, he ends up breaking with Sellars on the question of reductionism. What forces this is Pitt's embrace of the primacy of practice over theory. Unlike Sellars, who relies on a more linguistic conception of scientific realism, Pitt endorses a pragmatic criterion of scientific realism. That is to say, if one can reliably engage with an entity, it is real.[6]

To use an example from Hacking (1983, 265), because electrons can be used to affect other particles ("sprayed"), they are real. The same can be said for molecules, genetic material, neurons, and the "fundamental" forces of physics. All of these can both be manipulated and reliably so, especially with reference to other scientific entities. To ask questions of whether some scientific entity is real based only on theory is then problematic because of the shift of perspective. Instead of assuming that the world can be known without engagement with it, Pitt expects that there will be some sort of tie to practice somewhere. Pitt is quite fond of saying that "action is the mark of knowledge," which might be rephrased in this context as "acting with something is the mark of reality."[7] The ability to act with something, and more importantly, to use an entity to affect others, might not exhaust what counts as "real," but defines much of it.

TECHNOLOGY AND REALISM

What allows Pitt to defend this pragmatic criterion of scientific realism is his work on technology. Technology comes to play a decisive role in his thought in the 1980s, which marks yet another break between student and teacher. Like many philosophers of science before the New Experimentalists, and, really, many philosophers going back to Aristotle, technology was not on Sellars's radar. The assumption was, in the realms of truth and reality, that theory, and *not practice*, was where to look. Even from his first forays into linking technology to the philosophy of science, Pitt made clear that this assumption would not do. His 1983 article "The Epistemological Engine" demonstrated the role that instruments played within science, and that they were the *engine* of scientific discovery. The more he worked on this theme, the broader his definition of technology became until he reached the following: "technology is humanity at work" (Pitt 2000, 11. Emphasis removed). A brief consideration of this definition shows its scope: anywhere humans are at work appears to be technology, which means that perhaps everything humans do is technology.[8] This definition is deliberately broad and intentionally deflationary. One frequent criticism is, given this, what does not count as technology? Pitt is not fazed by this criticism since the whole point of the definition is to be broad and deflationary in such a way that global claims about technology, a la Heidegger, are impossible to make and philosophers of technology turn to the specifics of technolog*ies*.

With respect to the theory/practice distinction, the broadness of Pitt's definition becomes vitally important. Given how Pitt understands "humanity at work," and his discussion of the "technological infrastructure of science," the sorts of theories discussed above with respect to Sicilian realism (mature, well established, etc.) will count as technologies. Such theories are *tools*, supported by other tools (in a more conventional sense), for explaining, manipulating, and transforming the world. This, in effect, deconstructs the rather long and problematic association of technology with practice. Instead of placing theory in a position of superiority over practice, Pitt subsumes theory under the heading of practice.

What follows from this move is that *theory* should be regarded as one particular sort of practice, a sort of technology. Such a rejiggering of the standard arrangement of philosophical concepts begins to dislodge the long-standing tie between theories and claims about realism. Instead of simply assuming, for reasons primarily of tradition, that theories determine commitments about what is real and what is not, at a minimum it is now practice, understood broadly enough to encompass theory, that would be the arbiter of claims about reality. This shift should at least dislodge debates about reality from sentences and seminar rooms to laboratories and instruments. It is not simply a matter of what is said to be real in the theories of science, but rather what is done with those theories. Such doings might be establishing relationships between one theory and another, which is rather similar to the concerns of the traditional philosophy of science. What is more significant is what is done in the laboratory or the field with instruments, tools, other technologies—and in rich social and historical contexts and knowledge communities (if you extend Pitt's work to the reaches of engineering with the work of Ann Johnson and Walter Vincenti). The reason this is more significant is that it involves the manipulation of the world, rather than simply the manipulation of sentences.

Given this, it is doubtful that it is by coincidence that Pitt's best discussion of Sicilian realism comes in the final chapter of *Thinking about Technology*. The dominant theme of that chapter is what he refers to as "the technological infrastructure of science" (where his realism is almost an afterthought). The purpose of that chapter is to explain how technology, here understood not just as artifacts and instruments, but also social structure, provides a way of explaining scientific change. As he puts it: "[P]rogress in science is a direct function of increasing sophistication not merely in instrumentation, but in the technological infrastructure that underlies and makes mature science possible" (Pitt 2000, 123. Emphasis removed). The reason that technology plays such a decisive role in fueling the development of science is the way in which instruments and other technologies open new spaces for scientific research.

Pitt's perennial examples of this are telescopy and microscopy. Each of these opened up new possibilities for scientists to explore. First, light telescopes allowed for making new levels of detail visible to astronomers, then followed in the twentieth century by the emergence of other sorts of telescopes that relied on parts of the electromagnetic spectrum that are outside of human vision. Microscopes revealed worlds too small for the human eye to resolve. In both cases, the expansion of "vision" was not simply a matter of refining optics involving the visible light spectrum. Many other sciences and technologies feed into the increased sophistication of these means of extending sight. While microscopes were invented shortly after the telescope, their significance lagged behind telescopes until the development of industrial dyeing techniques because reliable stains were necessary in order to reveal the contents of most sorts of cells.[9] This technological infrastructure expands the realms in which scientists can engage with the world. Telescopes and microscopes provide just one example of how instruments allow for the perception of previously unobservable (or at least difficult to observe) entities. Beyond light telescopes, there are also those that work with electromagnetic radiation that is outside of human vision like those that detect X-rays or radio waves. There are also traditions of instrumentation that do not rely on a visual framework per se, but instead on hermeneutics (Ihde 1991); the meters physicists use to measure currents, scales to weigh samples, reactions used to identify chemical substances, and so on. Technologies expand the realm of perception of scientists, but not just for scientists. Hacking (1983) rightly points out that perception is a more active process than early modern philosophers tended to assume, especially perception through microscopes. Even if we ignore this, technology obviously allows for new ways for scientists to manipulate the world. A particle accelerator does not "merely" allow a physicist to perceive particles, but to smash them together and observe the results. So too with the various tools that molecular biologists use to manipulate genetic material. Perception, understood very broadly, is necessary in order to manipulate those things under scrutiny, but this is not perception for its own sake. Rather, it is with the end of manipulation in view. And it is technology that allows for all of this to occur, and the growth of the technological infrastructure of science allows for science to engage with new parts of the world.

From this point, it is a relatively short step to Pitt's expansive Sicilian realism. When a biologist seeks to isolate and determine the properties of a particular segment of genetic material, much of what they work with is taken as real. It need not be the properties of the genetic material itself that is initially taken as real, because determining whether those properties are real is the object of inquiry. But the entities and processes used to investigate will be. Furthermore, the biologists typically investigate and manipulate the material, not as an assemblage of atoms, but as genetic material that possesses certain properties that occur at that level. Whether the biologists actually worry about what occurs at levels lower than genetic material will depend largely on the results of inquiry. In this sense, the biologists work with the assumption that what they use to investigate the object of their inquiry is real. These tools are taken to be real. Depending on the results of the inquiry, the object of study might prove to be real as well. A good mark of this would be whether that entity or process can be used to manipulate and/or affect other entities or processes. The same can be said of the other sorts of scientific entities. When chemists work with molecules, the properties of electrons matter quite a bit, but the specifics about subtopic particles are largely, if not entirely, ignored because they are not relevant to the sorts of inquiry they engage in. So too with geologists, astronomers, anatomists, etc. They approach the entities they investigate not necessarily as assemblages of FEPs, but rather as entities appropriate to the level of the world they inquire into. Furthermore, scientists' capacity to work with, on, and through these entities reliably means that qualify as "real" in the sense discussed above.

Once debates about the reality of this or that scientific entity are moved into the laboratory (broadly construed), reductionism becomes less appealing. The assumption that reality is something fundamental and simple has a long history in philosophy, and Pitt never attempts to offer a knock-down-drag-out argument against this assumption. Instead, his approach to scientific realism in effect tries to displace the assumption. He seeks to show that we can do without. As discussed above, scientists working in the laboratory treat their various instruments, concepts, reagents, samples, etc. as real. These are taken as the stable reality that allows them to answer questions about what they do not understand yet. When in the midst of scientific inquiry, all of what allows the inquiry to be settled is taken as real. The reduction of all this to the FEPs occurs outside of that inquiry. The sort of realism that Sellars relies upon is an abstraction. Sellars's reductionist realism abstracts out certain features of the world like fundamentality and simplicity and then prioritizes those over all others. Within the context of certain forms of inquiry, this abstraction might be entirely called for, as is its prioritization over other possible abstractions. What is ruled out for Pitt is the prioritizing of this abstraction over everything else simpliciter. That finding methodological justification for this is challenging can be seen by thinking about technology. Given Pitt's rather broad definition of technology, a concept like "the scientifically real" is itself a technology. As noted above, Pitt allows for very few general statements that apply to all technologies. Particular technologies must be dealt with in their particularity. There is no one account of technology "to rule them all . . . and in darkness bind them" (Tolkien 1994, 49). It can prove impossible to specify once-and-for-all the function of a technology, given the tendency of humans to appropriate a technology designed for one function to others, but there can still be limits on what a technology can be used for (like I have yet to find a way to make my computer function as a microwave). If the abstraction that FEPs are *the* scientifically real is to be held as the only possible sort of scientific realism, the use of this technology in this way requires a justification. By shifting from talk about reference and sentences to technology shows this need. FEPs are one tool. But like any tool in a toolkit, it can only go so far. Its usefulness depends on the context of its use. That a given tool is not useful for a particular project does not make it any more or less real than those tools I rely on. FEPs have great utility for certain sorts of explanations and for investigations into certain sorts of other phenomena. In other circumstances, other scientific entities like molecules, genetic material, or galaxies prove more advantageous for inquiry. What marks all of these as "real" is that they are used to help settle questions about what is not yet understood.

We then find that the real is not a "given" but a "taken." Of course, for those who know Sellars's critique of "Givenness" will rightly note that the scientifically real is not a Given in the sense Sellars criticizes. There the Given is a concept used to put epistemology on a firm foundation because of its peculiar status. It is something brutish, indubitable. For anyone who has spent time in a laboratory or read the history of science knows how hard-fought the results of science are. They are not Given in the sense of "just being there." In one sense, something that is scientifically real is taken away from this work. Yet, the reason I liken Sellars's scientific realism to a Given is that, accepting that knowing about the real is the product of inquiry, in a metaphysical way it serves as a firm, indubitable foundation once it is epistemically established. Sicilian realism does understand the real as something of a starting point but in a more limited sense. The scientifically real is taken as indubitable not across the board, but in the context of a particular inquiry. That genetic material is composed of FEPs, as well as molecules, can be safely assumed when researching the genetics of a stretch of DNA, along with the basics of chemistry and so on. In this sense, their reality is a given, but not a Given. In light of further inquiry, something taken as real can be called into question. Given Pitt's commitments to both, broadly understood, pragmatic epistemology and the necessity of relying on mature scientific theories, what can be regarded as real *is not* the entity, object, process, etc., under investigation. Rather, it is also those scientific entities, objects, processes, etc., that support the inquiry into what is under investigation that are *taken* as real. Sometimes this taking might be unwarranted, as was the case in Galileo relying on a false optical theory when he sought to justify the validity of his telescopic observations (Pitt 2000, 5–6). To the extent that Pitt is a fallibilist, he will readily admit this. Clearly, more work is necessary to nail down when the accolade of "real" is licensed and what Sicilian realism entails metaphysically, but both of these concerns go well beyond the scope of this paper. Even with these caveats, I hope to have put Pitt's Sicilian realism on a slightly firmer footing.

TEACHERS AND STUDENTS

To conclude, return to the question raised at the beginning: how far can a student move beyond their teacher before they are no longer "a student" (except in the rather mundane sense of having a course with this or that professor or that they were a dissertation adviser or committee member)? What concerned Pitt early in his academic career came directly from Sellars, and I think it is safe to say that Pitt's journey to Sicilian realism began with this engagement. Even if Pitt abandoned Sellars's scientific realism and avoided the excesses of Sellars's writing "style," there is an important sense that Pitt's approach to philosophy is indelibly Sellarsian, not just Pitt's occasional endorsement of Sellars's definition of philosophy (Pitt 2011, xiv). Rather, Pitt learned how to philosophize from Sellars, like through the relentless use of dialectics to explore a position under scrutiny. Both seek to articulate their positions through the interrogation of their opponents, whether the opponent is real or fictional. Rarely is this used as an opportunity to simply tear down those with whom they disagree. Instead, these examinations work to bring out what is correct in their interlocutor's position, but disengage it from what is wrong. In this way, they draw together divergent concepts into a sometimes uneasy constellation. Although understood differently by each, both appreciate the necessity of a multiplicity of perspectives. It is this that might be the most significant legacy of Sellars for Pitt's thought (and mine as well). One way to make sense of Pitt's trajectory is that he began with Sellars's two perspectives—the manifest and scientific images—and ended up going beyond Sellarsianism through an attempt to take this perspectivalism more seriously. If this is accurate, then perhaps the distance between them is smaller and Pitt is in fact a genuine disciple of Sellars.

There is a further problem here as well. Pitt started where Sellars was, but ended up moving into a rather different conceptual space. Can the opposite happen? That is, beginning in a very different place and then ending up in agreement? Can that person be called a disciple of the master? Throughout graduate school, I fought against Pitt on any number of issues. Yet, since then, I find myself agreeing with him on a distressing number of points, but because of my engagement with philosophers he never approved of. In the case of this chapter most of my references are to either Pitt or to Sellars, but the inspiration for the argument owes much the classical pragmatism of Peirce, James, and Dewey. I learned a great deal about these pragmatists from Pitt and, as mentioned earlier, he would be quite comfortable being associated with philosophers. I expect that he would be less amused with the role that certain disreputable Continental philosophers played in its formulation (even if only indirectly. For example, it was not by coincidence that Derrida's reflections on Foucault introduce the chapter).[10] Here we find a case of a disciple relying on divergent voices, some appreciated at a distance and others frankly detested by Pitt, in order to make an argument in support of the master. Am I still a student? Can I still be said to be a disciple of my master? Perhaps yes, if for no other reason than we share a pleasure in perverting the thought of our respective masters.

NOTES

1. So, currently the most fundamental entities would be those proposed by the standard model, though they might be reduced to strings of some sort depending on the fate of that theory. Everything else is simply an aggregate of those truly real entities.

2. Unfortunately, two of the other places where he deals with this, "Sellarsian Anti-Foundationalism and Scientific Realism" (2010) and "Discovery, Telescopes, and Progress" (2011, 85–95), do little to advance his case further than what is found in *Thinking about Technology*.

3. Along with different levels, like those found in physics (macroscale, molecular, atomic, subatomic, etc.) or biology (ecosystems, bodies, cells, nuclei, etc.), different sciences can approach the same level using distinctive tools. For example, both physicists and chemists can work with quantum phenomenon, but approach the same entities (and the same equations) in distinctive ways. This phenomenon is more prevalent in the social sciences, where the same object of study can be approached using the tools of at least two different sciences. For example, an economic transaction could be studied via economics, psychology, sociology, and anthropology. A more complex treatment of Sicilian realism should address this phenomenon as well, but to do so goes beyond the scope of this chapter.

4. My dissertation attempted to make just this case. The committee was chaired by Pitt, and he neglected to disapprove of the argument there.

5. Of course, a less than charitable interpretation of Sellars would note he faces the very similar difficulty, but that he obscures it with so much Kantian claptrap.

6. Other qualifications are necessary but will be skipped over here. Specifically, given Pitt's concerns with astronomy, this definition requires refinements such that it can clearly account for stars and other celestial entities as real. In principle, this should be doable. Call this a Sellarsian "promissory note."

7. See, for example, Pitt (2011, 166).

8. While it follows that everything humans do is done through technologies, it does not follow that everything humans do is in fact a technology. One thing Pitt obscures in his own writings is the significance of play in human activity. Given Pitt's definition of technology, play should be something distinct from work, hence technology.

9. Pitt deals with precisely these sorts of cases in his forthcoming book *Heraclitus Redux: Technological Infrastructures and Scientific and Social Change*.

10. For those keeping score, the critique of Sellars's scientific realism as a form of Giveness is an appropriation of Nietzsche, especially analysis of nihilism in *The Will to Power*, and Derrida, like his "Structure, Sign, and Play" (Derrida 1978, 278–93). Both Marx and Heidegger's writings on praxical engagement with the world informed the discussion of "taking."

REFERENCES

Derrida, Jacques. 1978. *Writing and Difference*. Translated by Alan Bass. Chicago: University of Chicago Press.

Garnar, Andrew Wells. 2007. "An Essay Concerning Subjectivity and Scientific Realism: Some Fancies on Sellarsian Themes and Onto-Politics." PhD diss., Virginia Tech.

Hacking, Ian. 1983. *Representing and Intervening*. Cambridge: Cambridge University Press.

Ihde, Don. 1991. *Instrumental Realism: The Interface between Philosophy of Science and Philosophy of Technology*. Bloomington: Indiana University Press.

James, William. 1950. *The Principles of Psychology, Volume 1*. New York: Dover Publications.

James, William. 1987. *Writings, 1902–1910*. Edited by Bruce Kuklick. New York: Library of America.

Mitchell, Melanie. 2009. *Complexity: A Guided Tour*. Oxford: Oxford University Press.

Pitt, Joseph. 1983. "The Epistemological Engine." *Philosophica*. 32: 77–95.

Pitt, Joseph. 1997. "Developments in the Philosophy of Science 1965–1995" in *The Encyclopedia of Philosophy, Supplemental Volume*. New York: Macmillan.

Pitt, Joseph. 2000. *Thinking about Technology: Foundations of the Philosophy of Technology*. New York: Seven Bridges Press.

Pitt, Joseph. 2010. "Sellarsian Anti-Foundationalism and Scientific Realism" in *Self, Language, and World: Problems from Kant, Sellars, and Rosenberg*, edited by James R. O'Shea and Eric Rubenstein, pages 171–85. Atascadero: Ridgeview Publishing Company.

Pitt, Joseph. 2011. *Doing Philosophy of Technology: Essays in a Pragmatic Spirit*. London and New York: Springer.

Sellars, Wilfrid. 1991. *Science, Perception and Reality*. Atascadero: Ridgeview Publishing Company.

Sellars, Wilfrid and Paul Meehl. 1956. "The Concept of Emergence" in *Minnesota Studies in the Philosophy of Science*, Vol. 1, edited by Herbert Feigl and Michael Scriven, pages 239–52. Minneapolis: University of Minnesota Press.

Tolkien, J. R. R. 1994. *The Lord of the Rings*. Boston: Houghton Mifflin.

Chapter 4

Quasi-fictional Idealization

Nicholas Rescher

PRELIMINARIES

I feel confident that Joe Pitt's interest in pragmatism and physical technology will conjointly give him some degree of sympathy for an exercise in conceptual pragmatism such as the present discussion. For it is the task of the present chapter to argue that even certain fictions apparently disconnected from reality can prove useful as thought-instrumentalities for dealing informatively with "the real world." For just as physical technology finds its validation through its pragmatic efficacy in goal realization (as Joe rightly stresses) so conceptual devices—though artificial—can find validation in easing our way through the complex challenges for managing our real-world affairs.

THE GENERAL IDEA

There are a great many *real* apples on the fruit stand of the neighboring supermarket. And there are lots of *fictional* apples in the orchard of the Green Gables farm on Prince Edward Island. But in between the real and the fictional, there is also a category of apples that are neither the one nor yet quite the other. These would include such contrastively idealized things as *typical* or *ideal* Macintosh Apples.

And this is a situation which arises with respect to thing-kinds in general. For with such a kind Z, we can project the idea of a typical Z, a normal Z, an average Z, a perfect Z, or an optimal Z. And while we understand such items fairly well, we also realize that we may never actually encounter them in the real world. In this sense, there are things whose contrastive characterization within a group is a mode of quasi-fictional idealization. It is with such

Table 4.1 Kinds and Their Components

Grouping	Illustration
Kind	Canine
Sub-kind (species)	Sighthounds
Ultimate kind (lowest species)	Irish Wolfhound
Specific variety	Male adult Irish Wolfhound

potential existents that float indecisively between the actually real and the conjecturally fictional that the present discussion will be concerned.

MACHINERY

Some clarification of concepts is needed from the outset. We shall understand a *kind* of thing, as theorists have done since Greek antiquity, namely either as a natural kind (an Indian elephant) or as an artificial kind (a label on a golf ball). Kinds generally disaggregate into sub-kinds, ultimately into ultimate kinds (*infima species*), and then into varieties (see table 4.1).

Moreover, our deliberations will have to resort to a certain amount of expository formalism. We will need to resort to the following ideas:

- The conception of a classificatory *thing-kind* (Z, Z_1, Z_2, etc.) be it natural (corgis) or artificial (recliner chairs).
- The conception of a *contrastive qualifier* (Q, Q_1, Q_2, etc.) for items of a thing-kind, such as an *ordinary* Z, a *normal* Z, an *unusual* Z, an *average* Z, or the like.
- The conception of a *descriptive property* of an item in contrast to a *nondescriptive* (and frequently *evaluative*) property. (Being *lame* is a descriptive property of a dachshund, being *cute* or *lovable* is not.)

The descriptive properties of kind-members always indicate a determinable characteristic of some sort such as being snub-nosed, long-lived, or tawny-colored. The possession of such properties can, in general, be determined by inspection of the item at issue in isolation. By contrast, nondescriptive features cannot be determined from the descriptive constitution of the item. Examples of such features would be a vase's being "owned by Napoleon" or a painting's being "inspired by Renoir."

With many contrastive qualifiers for kind-members, the situation will turn out to be *quasi-fictional* in that the realization of the putative item is problematic—such a thing may or may not actually exist. For a given kind may simply fail to contain any typical or normal member whatsoever. The idea of such *Q*-mode *Z* may in the end prove to be an unrealized fiction. Let us consider such situations more closely.

TYPICAL

To implement the idea of a typical *Z*, let us take dogs as an example. There just is no typical dog as such, across the entire range from mastiff to chihuahua. If we want typicality, we have to go to something as specific as a mature female corgi. Accordingly, to capture the idea of typicality one would have to say something along the lines of the following specification:

> Given a taxonomically ultimate kind *Z*, a *typical* or *paradigmatic Z* would be one that (a) has all the descriptive properties that *Z*'s generally [i.e., almost always] have, and (b) has no descriptive properties that would mark it an eccentric "odd man out" among the *Z*'s (in that only very few of them have this property).

The need for the limitation to *descriptive* properties in this formula should be clear. Thus location, although a property unique to any given *t*, would not impede typicality since it is relational rather than descriptive.

NORMAL

A normal *Z* is one that (1) has *almost all* of the descriptive properties that *Z*'s almost always have and (2) lacks *almost all* those descriptive properties that *Z*'s almost always lack. Given this understanding of the matter, it follows that while a typical *Z* will always qualify as normal as well, the reverse is not the case. For present purposes this overly brief account will have to serve. However, there is a vast literature on statistical and operational normality, as well as social and behavioral norms.[1]

ORDINAL/USUAL

An ordinary (or run-of-the-mill) *Z* is one that tracks the statistical norms in that it has all of the descriptive properties that most *X*s possess and that lacks most of the properties that most *Z*s lack. However, it is theoretically possible that no actual *Z* whatsoever is ordinary—that is, that all the *Z*s there are non-ordinary in some way or other. More common of course would be that situation in which all of the *Z*s there are actually are of the usual sort. In that event, however, we will be able to say things like: If there were a *Z* that has the property *F* (e.g., "If there were a dachshund that has two tails"), it would be a most unusual one.

AVERAGE

Consider the idea of an *average* Macintosh apple. To qualify as such, the apple would need to be average in size, in weight, in sugar content, in the number of seeds, in red skin-coverage, and so on—with all measurable descriptive properties included. For it is of course more than likely that no actual apple would fit this bill across the board. Even when there are a great many *Z*s, it is easily possible that no average *Z* actually exists—if this is to call for averaging out in all measurable features.

And in fact they need not even always be possible. For even average *Z*s can prove to be impossible. Thus consider a group of three rectangles (p_1, p_2, p_3) answering to the description

	R_1	R_2	R_3	Average
Base	3	4	5	4
Height	3	2	4	3
Area	9	8	20	12 1/3

Note that in this particular group a triangle that is average in base and height cannot possibly also be average in area.

ON PERFECTION AND ITS EXCLUSIONS BY MERIT COMPLEMENTARITY

A perfect (or ideal) *Z* is one that has in maximal degree all of the positive features that can characterize *Z*s, and has in zero degree all of the negative features that might do so. Such things are certainly possible. Thus a mint can proudly claim to have produced perfect pennies all day long.

Indeed, it transpires that thing-kinds require perfection. (There are really no imperfect circles.) However, other thing-kinds by nature preclude it.

To illustrate this, consider the evaluation problem posed by the quest for a suitable dwelling. Here aspectival fission leaps to the fore. Consider such factors as size and placement. In point of size, spaciousness affords more room, is more impressive, more amenable to entertaining; but comportness small is more intimate, easier to maintain, less costly to heat, etc. And as regards placement, being in town is more accessible to work, shopping, whereas location in the suburbs means less traffic, more neighborliness, etc. Here and elsewhere, the various parameters of merit compete with one another in their being an evaluation.

Desideratum complementarity thus arises when two (or more) parameters of merit are linked (be it through a nature-imposed or a conceptually

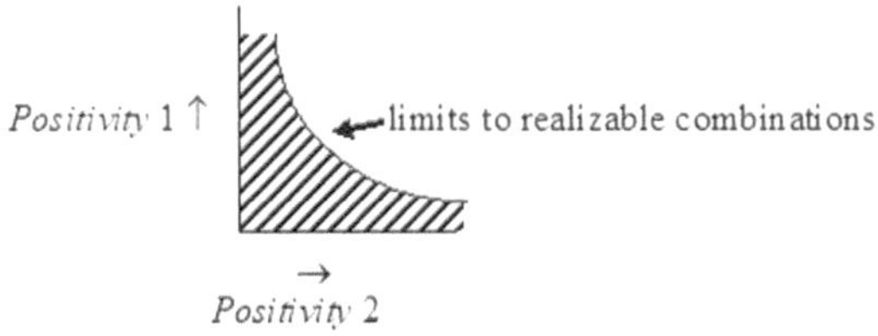

Figure 4.1 Desideratum complementarity. *Source*: Author created.

mandated interrelationship) in a seesaw or teeter-totter interconnection where more of the one automatically ensures less of the other, as per the situation of figure 4.1.

Throughout such complementarity cases we have the situation that, to all intents and purposes, realizing more of one desideratum entails a correlative decrease in the other. We cannot have it both ways, so that ideal of achieving the *absolute perfection* at issue with a concurrent maximization of every parameter of merit at one and the same time lies beyond our grasp as a matter of principle.

OPTIMAL

When the various instances of a kind *Z* differ in value, the existence of a uniquely best of them may nevertheless be unachievable for three very different reasons:

(1) There may possibly not be a single best because all of those *Z*s are just plain bad, with none of them any good at all. (There is no such thing as an optimal catastrophe.)
(2) There may possibly not be a single best because there is a plurality of them maximally qualified.
(3) There may possibly not be a single best because no matter how good some *Z* is there is always another yet better.

In all of these different ways, the goal of evaluative optimality may be unachievable within the range of alternatives afforded by a given kind *Z*.

All the same, there may be circumstances in which, even though no optimal *Z* exists, we may yet be able to say what an optimal *Z* would be like if there were one. A *Z*-type item which lacks to the greatest extent all of the shortcomings that actual *Z*s have would be indicated. Thus, while none of the students in a course produce an optimal result on the examination, an instructor can provide the graders with a model exam paper that realizes the desiderata of optimality—and will accordingly prove to be useful to them in their mission of grading the actual results.

THE USE OF IDEALIZATIONS

Each of the various quasi-idealizations at issue here, such as typical, normal, or average, instructs of kind-exemplification and relates to items which, for aught that we know to the contrary, may well not exist. The items at issue purport a descriptive specificity that can preclude its actual realization. At bottom, such things are creatures of thought (*entia rationis*) rather than furnishings of the real world. But all the same, even if there is no typical *Z*, no average *Z*, or no ordinary *Z*, the contemplation of such an item puts into possession information about the nature of reality. Even if no pupil in the class is an average performer, still knowing what such a pupil would be like provides valuable information. Irrespective of its realization, the nature of such a *T*-type *Z* conveys a good deal of information about the composition of the entire group of actual *Z*s at issue.

And so, despite their quasi-fictionality, such idealization can serve a useful communicative purpose in that they can function in ways that transmit useful information. We know a great deal about the animals at issue when we characterize a typical beagle or bulldog.

In sum, those quasi-fictional idealizations are a useful communicative resource able to render good service in descriptive and evaluative matters. The items at issue may not exist as such, but like the North Pole or the Equator, they are pragmatically useful thought-instrumentalities that facilitate our proceedings within the real world. Their idealization does not impede their utility in communicating about the real. Irrespective of their possible irreality as such, their employment can be highly instructive and constructive aids to understanding the world's actualities and dealing with them. In the final analysis the core idea of pragmatism is that the proper adequacy test of our technological and conceptual resources is their efficacy in our purposively oriented operations. And this core idea has figured throughout Joe Pitt's many-faceted pragmatic deliberations.

NOTE

1. A start on the subject is provided by Syristova (2010) *A Contribution to the Phenomenology of Human Normality in Modern Times*, Wikipedia as well as encyclopedias of philosophy, social sciences, and statistics constitute further useful resources.

REFERENCE

Syristova, Eva. 2010. *A Contribution to the Phenomenology of Human Normality in Modern Times*. Dordrecht: Springer.

Chapter 5

Technological Knowledge in Disability Design

Ashley Shew

The idea of technological knowledge—that is, know-how that cannot simply be put into sentences and evaluated—has been explored by scholars in philosophy of technology, philosophy of engineering, and in science and technology studies. In this chapter, I apply the idea of technological knowledge to my discussion on disability studies, especially about the expertise of disabled people when it comes to design. Through the work of historian of engineering Ann Johnson and the work of philosopher of technology Joseph C. Pitt, I highlight the communal nature of technological development and imagine a better frame for thinking about disability design, with and not just for disabled communities. In particular, the work of Pitt and Johnson asks us to consider communities of practice, the embeddedness of designers, and competing notions of expertise. For a long time, disabled people have been considered the recipients of technologies—not the creators or designers. Even when disabled people have been involved in design, their work is often seen in a helping role, rather than an intellectual, generative one. This is the wrong way to think about disability design if one hopes to make good things for disabled people. Technological knowledge is hands-on, knowledge-of-action-and-experience that cannot be arrived at by simulation or imagination.

INTRODUCTION

In this chapter, I explore the concept of technological knowledge as it is described by Joseph C. Pitt and by Ann Johnson. Ann Johnson writes:

> "Knowledge is produced in and by communities. . . . A knowledge community is a socio-epistemological structure, and I do not privilege either the social or

the epistemological dimension. As a result, I do not want to describe a body of knowledge independently of a community, implying that free-floating ideas and artifacts came together and attracted a community of practitioners" (Johnson 2009, 4, emphasis added).

Ann Johnson and Joe Pitt have, perhaps, opposite trajectories to arrive at the topic of technological knowledge but draw from similar sources and appreciate the importance of history and philosophy of technology done together. Johnson is a historian of technology and engineering who picks up on major themes and ideas, making philosophy relevant within the context of a practice. Pitt is a pragmatist philosopher of science and technology who takes historical context as crucial to not getting the philosophy (or the story) wrong. Over their careers, which have intersected at various points (they co-organized the Society for Philosophy and Technology biennial conference of 2007, for instance), they have each separately written about technological and engineering knowledge. In this chapter, I bring their work to bear on important and transformative work being done right now in disability studies and disability activism—around who gets to count as an expert when it comes to designing assistive technologies and other objects aimed at disabled bodies and minds (or, bodyminds[1]). Liz Jackson writes:

"As a disabled designer, I have come to believe that products are a manifestation of relationships. Disabled people have long been integral to design processes, though we're frequently viewed as 'inspiration' rather than active participants" (Liz Jackson 2018, emphasis added).

This chapter posits that what disabled activists are doing right now to elevate disabled expertise and ways of knowing coincides with historical-philosophical descriptions about expertise, community, and technological knowledge. In this way, I argue that the claims of disabled design activists about knowledge and expertise should be held in high regard—and these activists need to be heard. This chapter presents a prescriptive claim from an observational one. However, against a backdrop of a long history of ableism and nondisabled expertise about disabled people, understanding disabled people as offering expertise is an important and radical position to stake, hold, and proclaim.

Technological knowledge—often defined in opposition to scientific knowledge—describes knowledge that exists as know-how. Technological knowledge is knowledge of how to make and do. It is often contrasted against scientific or propositional knowledge: knowledge that can be expressed in words and declarations that can then be tested. Both Johnson and Pitt talk about engineering knowledge as a knowledge produced and used by engineers and tradespeople and as constituting a type of technological knowledge.

Both take tracing knowledge transformations as central to their work and appeal back to communities of design, the doers and makers, in their works on the topic. Expertise, trials, feedback loops, and materials all play a part in what can be established, known, and employed when it comes to technological knowledge.

DISABILITY ACTIVISM, SIMULATION, COMMUNITY, AND EXPERTISE

In recent years, disabled people[2] have been reclaiming their expertise with regard to disability design. Often viewed as the objects, beneficiaries, and recipients of design, with service projects abound and lots of rhetoric that are very negative about being disabled, disabled folks are speaking back against that view and asserting their own expertise and knowledge about their own lives. Liz Jackson, the founder of the Disabled List, a list of disabled designers available for consult and a group engaged in self-advocacy, writes in an article for the *New York Times*,

> "[O]ur unique experiences and insights enable us to use what's available to make things accessible. Yet, despite this history of creating elegant solutions for ourselves, our contributions are often overshadowed or misrepresented, favoring instead a story with a savior as its protagonist" (Jackson 2018).

Jackson details the story of OXO kitchen tools (a grippable kitchen utensil line) that are described by the company and others as the invention of Sam Farber, who is often described as developing the tools in trying to make kitchen tools that better fit the arthritic hands of his wife. In other words, the story is about a clever designer helping his poor, disabled wife. Jackson found her way to Betsey Farber, the unnamed wife with arthritis in this story to get her account:

> "The general understanding," Betsey told me, "was of the brilliance and kindness of Sam who made these tools for his poor crippled wife so she could function in the kitchen. I will probably go down in history as having arthritis rather than having the conceptual idea of making these comfortable for your hand."

Often, disabled people are conceptually involved in the creation of devices for disability: shaping designer knowledge and producing and explaining the knowledge that they have about the situation, disability, and device.

Even when not participating early in conceptually developing an idea about a design, disabled people serve as beta-testers and provide early feedback for different disability products (often unpaid and for virtually no recognition, in the case of many prosthetic devices and component designs as well as in the designs of other devices and infrastructures). For every good device for disabled people, there are disabled people in that device's history, shaping design decisions even when going unnamed and completely unrecognized as contributors, collaborators, or designers. This service to technological invention is often done for free, and companies that target disabled consumers often participate in exploitative labor that draws upon the expectation that all disability design is a form of charity for which disabled people should be grateful—and not compensated—to be involved with.[3] This notion runs contrary to respect for disabled expertise.

The most famous slogan from the civil rights movement for disabled people is this: Nothing About Us Without Us.[4] It is a hard-won and earned slogan for an often-disenfranchised and marginalized population. So many things have been determined and created "without us" disabled people to our detriment, exclusion, and death. A history where the "problem" of disability often translates to "the problem of disabled people existing" includes eugenics, sterilization, Ugly laws, institutionalization, audism, gas chambers, and involuntary euthanasia. The problems of actual disabled people can reflect the institutional biases built into our systems, as well as the ways in which disabled people are often spoken about and arbitrated without their own input. Mixed with racism, sexism, transmisia, and classism, community members experience ableism differently, and notions about disability (how disabled people are regarded) vary. But ableism also reflects deep prejudice about what constitutes a good life and who should be permitted to flourish.

Rarely are disabled people recognized, consulted, or approached as experts about disability—even more rarely are they paid for that expertise when we engage in the free consulting that disabled people are supposed to be grateful for. Poet and amputee cyborg Jillian Weise (2016) writes:

> I know it will take time, but things will change. For a while, all the experts on African-Americans were white. All the experts on lesbians were Richard von Krafft-Ebing. All the experts on cyborgs were noninterfaced humans.

Weise identifies cyborgs as interfaced humans—often humans with medical implants, dialysis machines, pacemakers, prosthetic devices, and other body-objects that often coincide with disability. Cyborgs—or disabled cyborgs, also called cripborgs (Nelson et al. 2019)—often experience the world with and through technologies that become sometimes seamless in experience and are sometimes made focal or noticeable with their malfunction or clunkiness.

Designer and Disabled List contributor Josh Halstead (2019) has provided a short definition of design that will appeal to some philosophers of technology: "Definition of design: A process by which human faculties and intentions *extend* into the life-world for further realization." Design, in and alongside disability, with disabled people, by disabled people, often results in technologies and systems better for all people. This is one of the calls around the push for universal design. Design "for" disability (notice: instead of design with disabled people), they say, and you make better objects for all. However, this push often undercuts the lived experience of disability within design. Instead, people are asked to empathize with disabled people and imagine themselves as if they were disabled—instead of actually consulting with or listening to disabled people.

There is a growing body of literature on the problems of disability simulation when it comes to understanding the experiences of disabled people. Disability simulations were popular exercises in "awareness" during the past few decades. People would don blindfolds or goggles to complete tasks to mimic blindness and low vision. People would use a wheelchair for a day or for a series of obstacles to imitate what it is like to be a wheelchair user. Michelle Nario-Redmond et al. (2017) writes:

> The primary goal when administering disability simulations is to grant non-disabled people an opportunity to improve understanding and acceptance of people with disabilities. Instead of just imagining what disability must be like, simulations allow people to role-play through personal experience. This kind of perspective taking is built on the assumption that people cannot fully understand the circumstances facing disabled people unless they know first-hand how disabled people seem to do what they do. (p. 325)

Nario-Redmond and her team explain that these simulations often engage in stereotypes. In the context of engineering and design, these simulations end up harming the community. Liz Jackson, this time in a talk called "Empathy Reifies Disability Stigmas," explains:

> The world has taught us that a disabled body is nothing more than a body in need of intervention, and this is what designers do, right? We fix things. . . . There's a certain glory in being a designer.

Jackson here tells a story about someone who teaches at a design school who dreads the start of semester where students come by wanting to fix or redesign her walker—a walker she likes. She resents being told there is something wrong with it. She continues in the same talk:

> We, disabled people, have become a topic. There is no rigor. We are nothing more than your portfolio enhancer or your brand enhancer. We are a symbol of your altruism. And that, among so many reasons, is why the very systems that would be in place to support us end up doing more harm than good.

Cassandra McCall and I are currently writing a paper where we suggest that it's time to "abandon empathy" in engineering education—because the way empathy in design and engineering is currently operationalized actually works against the people the designers empathize with. Simulation allows people to think they understand a situation without real-life experience in it. They pick the wrong problems and issues because what they have experienced is more akin to what it is like to experience a new state of body or mind, not what it is like to live in that bodymind every day. Simulation creates a tragic situation where nondisabled experts continue to reinforce their expertise and stereotyped biases about disabled bodyminds without having disabled bodyminds themselves, rather than listening to those for whom they design.

Experiential knowledge matters here—rather than what is simulated or imagined or empathized. So often disability simulation relies on the idea that disabled people are not worth consulting and that the experience of disability can give a person the phenomenological authority to author, design, or create on behalf of someone else. It also gets people thinking individually about their needs, rather than thinking of themselves as members of a community. The problem with empathy here brings us back to technological knowledge. Technological knowledge is deeply embedded in practice and experience. One cannot learn it as rote. Being deeply involved in the systems and community matters to the development of technological knowledge as well as the success of design.

TECHNOLOGICAL KNOWLEDGE

Technological knowledge is knowledge embedded in technology, and knowledge of how to make effective action with a technology. Knowing how to be effective is often a social process. Ann Johnson, in writing about the concept of reliability, explains how our work (in STS, as scholars of technology studies in some way) is oriented toward both technical and social explanation.

> As historians and sociologists of technology, we often unpack the thorny issue of what makes a given product successful, and in doing so, face the problem of striking a difficult balance between technical and social determinants. The success or failure of most new technologies cannot be explained strictly by looking at their technical attributes and the engineering criteria they were designed to meet. Nor can we usually explain a product's success by focusing solely on the social relations and arrangements that led to and accompanied its introduction. Instead, we must map out an array of factors from social to technical

to economic that led to the success of one technology over its competitors. (Johnson 2001, 249)

To make sense of successful design is to take technical and social factors together. This balance helps us make sense of prominent case studies of success and failure in disability design. Liz Jackson has coined the term "disability dongle" to describe a disability design without a realistic disabled public for which it was designed (see "A Community Response to a #DisabilityDongle"). Often a focus on technical aspects alone with a stereotyped idea of what constitutes a disabled body—so many dongles developed around an assumption that all wheelchair users are paralyzed (they aren't) might hype exoskeletons as a "solution" for an unclear problem. Without social engagement in design, by which I mean during design and not in adoption/non-adoption after designed, many disability products are utter failures.

What is it then that disabled people know that helps in design, helps in creating more universally desirable objects (for disability design in concordance with universal design here)? How do technologies come to embody and constitute a form of knowledge in the disability community? (Is this where we have cyborg studies and disability studies coherent within a philosophy of technology?)

Both Joseph C. Pitt and Ann Johnson write mostly of technological knowledge in terms of engineering knowledge. I work more broadly here because many designers and would-be designers and engineers from the disability community have experienced ableism such that accessing higher education, and especially rigorous (here read: ableist, inflexible, hostile) engineering programs, has traditionally been made much more difficult. Therefore, I write here about technological knowledge, a category I take as broader than engineering knowledge in terms of who can participate and who is considered prominent in this interaction with objects; engineering knowledge here is a subset of technological knowledge. Pitt defines technology as "humanity at work" in *Thinking about Technology* (2000). To talk about technology as knowledge, then, is to talk about the activities of humanity, and never just one human alone, and to see it as alive at work. Though I have taken significant issue with this definition elsewhere (see Shew 2017), my objections have been about limiting technology as the purview of humans only. But to think about groups of creatures, human or not, producing or creating in collaboration, at work together in community, gives me less hesitation.

When talking about the knowledge of science and technology, a distinction is often made, in the vein of Gilbert Ryle, between "knowing how" (technological knowledge) and "knowing that" (scientific knowledge) (Pitt 2011, 168). Riding a bicycle serves as an example here: even if you understand the mechanics of riding a bicycle (knowing that a bicycle works in a particular way, being able to give a description), you still don't know how to ride one from that description alone. The "knowing" that a person does around riding a bicycle involves embodied experience that cannot be reduced to or made into text. One learns how to ride—the feel of the bike underneath you, a sense of balance on it—all the steps to learning from perhaps training wheels or a tricycle or balance bike to more traditional forms. Of course, not all bikes are the same, so switching between, when possible, may involve learning how all over again—from the different feels of mountain bikes and beach/sand bicycles to ones meant for city riding. Embodiment plays a large role in bicycle-riding.

Bicycles also serve as an interesting example here because of all the different adaptations and adapted bikes that serve various communities: with crank-arm shorteners and Orthopedals for people with knee injuries and replacements, or lower range of motion aside from injury. Hand-crank bikes can be used to pedal with arms instead of legs, and adult tricycles serve well for people with brain injuries and people at risk of falls. Knowing how, when it comes to bicycles, is not one set of knowledge, but of self-knowledge with a technology. Bicycling, like so many other human-technology interactions, requires knowledge embodied by individuals and also by various configurations suited to particular environmental or built infrastructure niches—of place, of surface, and of how bodies are configured. Bicycles are "fit" or configured for riders with adjustments to seats and handlebars, even outside disability concerns. Surfaces matter to the bike's configuration—fat tires for sand and surf, much narrower for mountain biking, and road biking still different. Creating an effective bike for an effective rider involves knowledge about surface considerations, body configurations, and sometimes even weather. The task or goal—bike riding—is never just the activity alone.

Walter Vincenti explains, using engineering knowledge, that "[e]ngineers use knowledge primarily to design, produce, and operate artifacts" (1990, 226). He contrasts this with scientific knowledge used to produce accounts or produce more knowledge. Edwin Layton and A. R. Hall speak of technological knowledge as knowledge of "how to do or make things" (Layton 1987, 603). Pitt, explaining Vincenti, says that "[e]ngineering knowledge is task-specific and aims at the production of an artifact to serve a predetermined purpose" (2011, 170). Pitt repeats this theme: a task-orientation. He says this about architectural design as well. A goal orientation matters to how technological knowledge forms. There is some action needed to be done. This sends us right back to considering the social mechanisms that exist alongside the technical, to use Johnson's concepts.

For many disabled people, technologies are connected to functionality in our lives. The "major life activities" described in the Americans with

Disabilities Act, defining who counts as disabled against whether these can be done, for employment and accommodations, include (but are not limited to) caring for oneself, performing manual tasks, seeing, hearing, eating, sleeping, walking, standing, lifting, bending, speaking, breathing, learning, reading, concentrating, thinking, communicating, and working. To be unable to do one of these things without assistance (often technological) puts one under the legal definition of disabled. For me, I cannot walk without assistive technology—often my prosthetic leg, but I could also employ a wheelchair or a walker. Though major life activities may be "shot through with technology," in the words of Andrew Garnar, the major life activities of disabled people often witness a closeness to technology unlike the nondisabled: from our living and communicating: from feeding tubes and ventilators, to crutches and canes, to specialized keyboards and text-to-speech software, to fidget spinners and hearing aids, to knee replacements and ostomy bags.

Oftentimes technologies are also part of whether we are allowed or accepted or included or welcomed: ramps, seating, marked pathways, elevators, good acoustics, nonfluorescent lighting, navigable distances, materials in multiple formats, snacks uncontaminated by allergens, doors that are openable with ease or by button, places that aren't too busy with noise, no flashing lights or loud booms. Extra attention to designed spaces and designed objects is near to the experience of disability. For many people, to become or to be disabled is to have a keener sense of what is allowed and what is impermissible within a space and through a device. Some of this is more purely social: the push to pass as nondisabled looms large. Even if a space permits a person to lay on the floor for disability reasons, there may still be a huge incentive not to do so and risks to appearing disabled within a space (Herdegen 2019). As Pitt writes in *Doing Philosophy of Technology* (2011) about architectural design specifically: "There are spaces that we use for living, working, recreation, etc. Sometimes they contribute significantly to achieving the goals we seek to accomplish in those spaces and sometimes they do not" (p. 148).

Both Ann Johnson and Pitt credit knowledge communities to local conditions: for groups of knowledge experts to arise and work on projects together requires some local prompt to bring them together. This act places technologies in a relationship to ethics and history in very real ways. The story, then, about technological knowledge is that such knowledge rests within communities and not just within bodies. We may spell out local conditions, individual conditions to address particular people—with modification and adaptation, but knowledge about disability. This is embodied, yes, but also learned through experience, community, feedback, and trials, and it becomes the basis for good technological design. Technological knowledge in disability design needs disabled people. Even for nondisabled designers, design for disability requires engagement with a community of people.

I think here of my prosthetist's mode of working with the amputees he serves. He's not an amputee, but he trained as an engineer before going back to school and getting certified in orthotics and prosthetics. He approaches each day as an engineer. The first conversation an amputee has with him about goals we have—what tasks should we be oriented to in our design—and I say our design because the design is an iterative process, rife with feedback loops between the two of us as well as his prosthetic technician and many times also suppliers of parts for the leg. He pulls from a wide range of knowledge about components and materials—knowledge I do not have. But I know how things feel and walk and rub, and I work with him to suggest slight modifications or suggest how I would like it to be in some way. We test out different things along the way, too, with many small adjustments as a normal part of this process.

Small changes here matter too: when my second leg was built, I kept snapping rivets off the metal sides that flank my thigh. I had never had that problem on my first leg, whose rivets are still perfectly intact today (as the leg sits in my closet). The rivet manufacturer had changed where or how they were producing the aluminum rivets. I took more than a few emergency trips to hardware stores to get small screws and bolts to make sure what was once riveted together could stay together long enough before seeing my prosthetist again. (He works four hours away from where I am.) He eventually ordered copper rivets and replaced all the rivets on the leg, and I have had no more rivet issues.

And, of course, getting a new leg made involves not just my prosthetist and me, but also a large network of people developing components and materials. To walk with my new leg, my prosthetist draws on his time with physical therapists and with other patients in directing me in how to move and walk. He is willing to talk with a patient's local physical therapist to give them advice. This description is in the absence of talking about an even larger network of insurance and billing companies, surgical offices (that write prescriptions for legs and arms), and other various assistive technologies that one might use as a leg amputee (crutches, canes, walkers, and wheelchairs, but also grab bars, shower stools, adaptive automobile equipment). We might also consider systems of maintenance, social stigma, and taint that a person may encounter, and aesthetic dimensions of design. Knowledge about prosthetic devices is not solely located in any one of these areas, and the engineering knowledge at use in my prosthetist's office in the making of my device is accumulated and with no one originator. And the lived experiences with these technologies and components have deeply shaped how they exist today, knowledge instantiated into the technologies we have and use: rivets and Velcros and carbon fiber and their configuration as they hang off my body are not just about my body. There is no design from nowhere.

When Pitt writes, "technologies don't operate in a vacuum, they are always socially, geographically, and historically contextualized and what the context is makes a huge difference on which normative conclusions we draw" (Pitt 2011, xii, my emphasis), I think of how "engineering for good" and humanistic engineering programs currently frame disability design—and present themselves as designers looking for a problem without real attachment to the experience, context, and history that would make their work more complicated. This is reflected, for instance, in the excitement and investment around exoskeletons—devices which many wheelchair users reject or see as far less useful than, say, better research about bladder and bowel issues or interest in enforcing of current disability accessibility laws that would improve quality of life.[5]

One of my favorite things Ann Johnson wrote is a comprehensive book review where she looked at three different new books and where she "revisits" technology as knowledge. These books all help illustrate the ways in which engineers alone cannot be the only ones engaged in the making of various materials and systems and highlight the roles of training, tradespeople and artisans, labor, and commercial interests on how we understand technology as knowledge. Johnson writes, bringing together the themes of the three books she reviews:

> Taken together these three books provide an account of the current state of technology as knowledge. All three authors are concerned with this perspective, as seen by their efforts to show the nature of the knowledge communities in which their actors operated, how those communities facilitated the exchange of knowledge, and the social relations of their specialized communities with industry, labor and society more generally. (Johnson 2005, 564)

Johnson, through this review, highlights consensus about what constitutes technology as knowledge—and the result is a picture of technological knowledge that is highly community-based and involves different communities working out their configuration while setting standards and norms about how to build, make, or produce. The image of Pitt's feedback loops—between and through these sometimes competing communities—comes to mind.

Johnson highlights a number of themes within the three historical volumes she is reviewing and ties them to the work of philosophers and historians of technology in prior discussions of technological knowledge. She highlights:

- objectivity—as constructed and a "cultural quality," but that which also lends authority to particular knowledge communities;
- commodification—"commodification implies that this knowledge can be bought and sold" (p. 565), so commodification is also part of stories about technological knowledge;

- tacit technological knowledge—skilled knowledge, or knowledge about making, or embodied knowledge, is part of understanding technology as knowledge; this theme, underexplored in the literature, is "one determining factor in the spread" (p. 568) of the technologies in the books Johnson is reviewing; and
- materiality—the primacy of materials in creating new technologies and systems; Johnson explains materiality: "The role of engineers or of anyone who designs technical artifacts is to figure out the relation between physical and functional characteristics. Materiality is therefore a defining characteristic of technological knowledge" (p. 569).

Johnson also highlights two more interrelated themes: the authority and ethics of technological knowledge. Authority is related to objectivity, but is also part of the steady working of technological knowledge. Authority creates or reinforces particular social arrangements and hierarchies that lead to responsibilities on the part of many knowledge communities, requiring codes of ethics and reflection. I think of these categories that Johnson makes clear, drawing from so many sources, in relation to the ways in which disability community leaders are working.

TECHNOLOGICAL KNOWLEDGE IN DISABILITY COMMUNITY

Objectivity, Reworked

Authentic disability community leaders work against the notion that non-disabled people get to be the "objective" experts about disability. For many of us, we are categorized and understood primarily through a medical gaze that overlooks personhood, individuality, and agency. The move to promote the social and other alternative models of disability is one that counters this medical model that would see disability—see disabled people—as needing to be fixed, cured, monitored closely, or institutionalized. I see disabled activists and scholars countering a narrative that puts "helpful" doctors, therapists, and officials in charge of disabled people—denying that they were ever objective and reestablishing disabled people as the proper authority on disability. I direct readers to the title of one edited volume by autistic adults with advice for parents of autistic kids: *The Real Experts* (Sutton 2015).

Commodification

Disabled leaders already recognize that disability service and disability technology is big business, while disabled people are also more likely to be poor and more likely to be unemployed. The goal of leadership in this area is to get recognition for disabled invention, creativity, and making, with gestures toward appropriate compensation for much of the unpaid work disabled people do. There are movements around social safety nets and universal basic income, but the things made for/with disabled people are often sold widely without recognition. I think here of how weighted blankets have become commercially available while people who made early designs and served the community have been overrun by these larger distributors who market them to nondisabled people too. I think also of how folks like Liz Jackson and the Disabled List are approaching design communities directly for intervention and working to change museum curation practices around disability tech. The larger work of shifting narratives about disability and technology is one that is about how disabled lives are commodified, marketed, and understood.[6]

Tacit Technological Knowledge

Part of how disabled people are countering the narratives we see about our consumption and needs—both framed through medical models and part of how things are marketed—is to talk about the knowledge of experience of body with technology that cannot be captured by empathy or pretending to be disabled. Disabled people are claiming knowledge that can't be extracted and distilled for the interests of the nondisabled. This knowledge is about our myriad embodied experiences and our experiences with technologies, environments, and systems. This is part of reclaiming authority: we have a special knowledge, from our bodies, our community sharing, and our positionality. The disability community and those people working in it have far more knowledge about disability than individual disabled or nondisabled people, who may draw primarily from their own personal circumstance or from dominant tropes about disability in our culture.

Materiality

What disabled people know about technology is how to, in the face of physical and infrastructural limits, hack it, modify it, groupthink it, move it, network it, share it, and bend it. Leah Lakshmi Piepzna-Samarasinha writes of how disabled users work in a particular online group around material objects:

> SDQ [a facebook group of thousands of sick and disabled queer] folks regularly mailed each other meds and extra inhalers and adaptive equipment. We shared, when asked, information about what treatments worked for us and what didn't and tips for winning a disability hearing. We crowdsourced money for folks who needed to replace stolen wheelchairs, detox their houses, get living

> expenses together for rehab, or get out of unsafe housing situations. People sent care packages and organized visit teams for members they might never have met in person who were in the hospital, rehab or the psych ward. We cocreated an evolving, amazing cross disability best practices/community guidelines document that helps folks learn about the disabilities that aren't ours—from captioning videos to neurodivergent communication styles. (2018, 61)

What Piepzna-Samarasinha writes about above is about play with materiality and infrastructure from lived community disability experience, borne out of necessity. The material bits of our lives—some of which enable us to live at all—shape our experiences, understanding, and the things we both come to know—from this very material experience of the world—and come to shape.

Disabled people have a different material knowledge of the world that informs our collective experiences; and our experiences of individual technologies inform how they could be used, adapted, and hacked. There are many individual technology instances, and share our "lifehacks," in the language of Liz Jackson, that reflect this deep material and community knowledge.

Here I have a quote from Joseph C. Pitt, but rewritten to replace engineering and engineers with disability and disabled people:

> Because it is task oriented, and because real world tasks have a variety of contingencies to meet—e.g., materials, time frame, budget, etc., we know when a disability project is successful or not. . . . What disabled people know, therefore, is how to get the job done—primarily because they know what the job is. (2011, 174, with modifications italicized)

Disabled people know what the job is when it comes to things related to disability. Of authority and ethics, there are conclusions to draw about how our material world could be reconstituted by the elevation of the disability community as an authority on technological knowledge, from individual design to large structures. This has implications for how projects about disability play out, as well as serious re-envisioning of the ways in which everything from homes and transport to businesses and institutions is designed.

AFTERWORD AND ACKNOWLEDGMENTS

Joseph C. Pitt and I have both become disabled during our colleagueship. He took me on as a graduate student in 2005, directed both my master's theses (2007 and 2008) and my PhD (2011). I became disabled in 2013–14 due to osteosarcoma; I am a hard-of-hearing, chemobrained amputee. Through this time, Joe has had a number of the discs in his back fused and been in a car accident that had some significant recovery time. We started wearing hearing aids around the same time, and, about the time I stopped walking with a cane, Joe started using one. His is a gnarly old branch of a cane, and much cooler than the basic one I was sporting. We now participate together on the board of our local New River Valley Disability Resource Center. The NRVDRC is a center for independent living, part of a national movement started in the 1970s to include disabled people in decisions on, well, disabled life things. Part of the mission is to keep people who want to live in their own homes there, to transition people out of institutions, to support home modifications for those that want to stay in place, and to give information, referrals, and peer support. The constitution of the board for all centers for independent living has to be majority disabled people; they get the meaning of Nothing About Us Without Us . . . we get the meaning of Nothing About Us Without Us.

Unbeknownst to Joe, perhaps, is how I see his work on technological knowledge and communities as part of this work we do together at the non-profit NRVDRC. We know things and have relevant experience to do good work with this center. I owe a great deal of thanks to Joe for a good many things, both academic and personal. (So, thanks, Joe.)

Similar thanks are deserved to Ann Johnson, who was my undergraduate honors thesis adviser (2005) and an amazing mentor and confidante, especially as I went through a year of chemotherapy and surgery. I wish she were still with us so that I could thank her for so many things—including mentorship, camaraderie, and silly cartoons during dark days.

I also want to acknowledge the people and team I have the pleasure of working with at Virginia Tech. Talking over ideas about disability tropes, community, narrative accounts, and disability technology with this group is truly a pleasure and privilege—and one that has informed what I have written here. Particular thanks go to Joshua Earle, Hanna Herdegen, Damien P. Williams, Adrian Ridings, Stephanie Arnold, Kelly Dunlap, Martina Svyantek, Cora Olson, and Monique Dufour.

Note: This material is based upon work supported by the National Science Foundation under Award No. 1750260. Any opinions, findings, and conclusions or recommendations expressed in this material are those of the author(s) and do not necessarily reflect the views of the National Science Foundation.

NOTES

1. Margaret Price uses this term in many publications, including *Mad at School*, and many disability study scholars have supported this word as a useful expression. The word indicates that our bodies and minds are inextricable and that what affects a

body also does a mind, and vice versa. John Dewey also uses a similar term "body-mind" as early as 1925. Thanks to Andrew Garnar for pointing this out!

2. I use the term "disabled people," but recognize that there are debates within disability communities about person-first versus identity-first language. I prefer identity-first; I am a disabled person. Lydia X. Z. Brown has some wonderful and widely shared reflections on the matter (2011), and Beth Haller has a helpful resource for journalists navigating language around disability (2016). Most of the designers and activists in this movement that I am talking about identify as disabled people, so this choice makes good sense in this context.

3. Given the cost of different components—where designers and manufacturers charge a hand and a leg for a fake foot—it's absurd to think of design for disability as charitable. My last foot—a noncomputerized carbon-fiber shenanigan with a molded plastic outer—cost my insurance $4,000. That's without the rest of the leg attached.

4. Nothing About Us Without Us is a great slogan, but I'm also a little partial to PISS ON PITY.

5. *New Mobility*'s Crip Buzz over exoskeletons, one such example: http://www.newmobility.com/2014/08/crip-buzz-august-2014/.

6. The definition of who counts as disabled was shaped and developed under conditions of slavery, colonialism, and industrialization; what we value is often tied to productivity and work. Questions of who can work—which bodies and minds are fit for labor, are deserving of jobs and rights—have informed much of how the disability community is understood in our culture. Patty Berne, as quoted in Piepzna-Samarasinha (2018): "Disabled people of the global majority—black and brown people—share common ground confronting and subverting colonial powers in our struggle for life and justice. There has always been resistance to all forms of oppression, as we know through our bones that there have simultaneously been disabled people visioning a world where we flourish, that values and celebrates us in all our myriad beauty" (pp. 20–21).

BIBLIOGRAPHY

Brown, Lydia X. Z. 2011. "The Significance of Semantics: Person-First Language: Why It Matters." [Blog] *Autistic Hoya*, August 4, 2011. https://www.autistichoya.com/2011/08/significance-of-semantics-person-first.html.

Eames, Robin. 2018. Twitter Post. September 10, 2018, 8:34pm. https://twitter.com/robinmarceline/status/1039311147738906624.

Haller, Beth. 2016. "Journalists Should Learn to Carefully Traverse a Variety of Disability Terminology." *National Center on Disability and Journalism*, January 20, 2016. https://ncdj.org/2016/01/journalists-should-learn-to-carefully-traverse-a-variety-of-disability-terminology/.

Halstead, Josh. 2019. Twitter Post. February 22, 2019, 9:30am. https://twitter.com/Josh_A_Halstead/status/1098953283660464128.

Herdegen, Hanna. 2019. "Maintaining Disabled Bodies and Identities: Disability As Dirty Work." [Blog] *The Maintainers*, June 17, 2019. http://themaintainers.org/blog/2019/6/17/maintaining-disabled-bodies-and-identities-disability-as-dirty-work.

Jackson, Liz. 2018. "We Are The Original Lifehackers." *New York Times*, May 30, 2018. https://www.nytimes.com/2018/05/30/opinion/disability-design-lifehacks.html.

Jackson, Liz. 2019. "A Community Response to a #DisabilityDongle." *Medium*, April 22, 2019. https://medium.com/@eejackson/a-community-response-to-a-disabilitydongle-d0a37703d7c2.

Johnson, Ann. 2001. "Unpacking Reliability: The Success of Robert Bosch, GMBH in Constructing Antilock Braking Systems as Reliable Products." *History and Technology*, 17: 249–70.

Johnson, Ann. 2009. *Hitting The Brakes: Engineering Design and the Production of Knowledge*. Duke University Press.

Layton, Edwin. 1987. "Through the Looking Glass, or News from Lake Mirror Image." *Technology and Culture* 28(3): 594–607.

Nario-Redmond, M. R., Gospodinov, D., and Cobb, A. 2017. "Crip for a Day: The Unintended Negative Consequences of Disability Simulations." *Rehabilitative Psychology* 62(3): 324–33.

Piepzna-Samarasinha, Leah Lakshmi. 2018. *Carework: Dreaming Disability Justice*. Arsenal Pulp Press.

Pitt, Joseph C. 2000. *Thinking About Technology: Foundations of the Philosophy of Technology*. Seven Bridges Press.

Pitt, Joseph C. 2011. *Doing Philosophy of Technology: Essays in a Pragmatist Spirit*. Springer.

Price, Margaret. *Mad at School: Rhetorics of Mental Disability and Academic Life*. University of Michigan Press.

Shew, Ashley. 2017. *Animal Constructions and Technological Knowledge*. Lexington Books.

Sutton, Michelle, ed. 2015. *The Real Experts: Readings for Parents of Autistic Children*. Autonomous Press.

Vincenti, Walter G. 1990. *What Engineers Know And How They Know It*. Johns Hopkins University Press.

Chapter 6

The Effects of Social Networking Sites on Critical Self-Reflection

Ivan Guajardo

Marshall McLuhan writes in *Understanding Media*: "Technical change alters not only habits of life, but patterns of thought and valuation" (McLuhan 2001, 63). The debate between those who think of technologies as encapsulating values and those like Joseph C. Pitt, who think of technologies as value neutral is worth revisiting in light of the challenges posed by recent technologies like the Internet, Artificial Intelligence, and Genetic Engineering. On the one hand, some past and present scholars claim that technologies have normative properties, but disagree substantively about the meaning of this claim (Ellul 2003; Feenberg 1999; Winner 2000). Determinists like Jacques Ellul (2003) interpret it to mean that modern technology is a pervasive force that follows its own inner logic or set of imperatives, thereby imposing fixed patterns on society. Those who reject this strong form of technological determinism insist nevertheless that technologies often encapsulate values of a particular stripe, which they can reproduce by imposing certain requirements on users or by the effects their implementation can have on the world. On the other hand, those like Pitt who consider technologies—whether manufactured objects, applied knowledge, methods, or the sociotechnical systems they are part of—to be neutral, think of tools as *in and of themselves* impartial or unbiased with respect to particular values or norms (Pitt 2000, 2011; Bunge 2003).

This chapter brings to bear recent empirical and conceptual work on social networking sites (henceforth "SNS"), and their negative impacts on the psyche, on the disagreement between neutralists and non-neutralists concerning the place of values in technology. It begins by describing and raising objections to Pitt's arguments for neutrality while also identifying important elements of his philosophical work on technology that should be part of any critical evaluation of particular technologies. The case of SNS is then introduced, including a description of the values they embody—an ethic of personalization SNS maintain by generating conditions like epistemic bubbles that reproduce it. To meet Pitt's demand for assessments of technologies grounded on empirical evidence, the chapter describes, in turn, various studies linking SNS to epistemic bubbles and the variety of negative impacts these environments can have, like polarization. A hitherto overlooked but important way the patterns observed in these studies replicate the ethic of personalization is by disrupting people's capacity to moderately respond to reasons or "MRR," a necessary condition for guidance control of thinking and action, thus of critical self-reflection. After defining MRR, the essay catalogs ways the epistemic bubbles that personalized SNS generate disrupt it. To address technological problems like those being caused by personalized SNS, Pitt recommends communal, self-correcting, commonsensical design processes and teaching practices aiming at improving individual and collective technological decisions by learning from mistakes. The last section discusses this common-sense approach to technological problems. I argue then that Pitt's recommendations must be part of any sound approach to technological problems but not without taking into account that technologies can propagate the specific values embodied in their design in their own right by means of their consequences. This means that the communal, self-correcting, commonsensical design process Pitt recommends as the way to deal with technological problems must foster awareness of the specific values a given technology might come to embody, and how they may serve or fail to serve broader ends and values the community takes itself to be committed to. Fostering this awareness requires a collective effort to teach designers, owners, and users ways to recognize and evaluate the often subtle and invisible effects a technology's normative properties can have on individuals and societies. Pitt's approach must also be placed within a larger critique of how the political economy within which technologies exist relates to the specific values they embody and what this relation teaches concerning possible solutions to the problems empirical research connects with the use of these technologies. The conclusion considers how Pitt's common sense approach to technological problems and the supplementations that must be added bear on some proposed ways to redesign personalized SNS currently being discussed by the larger culture.

PITT ON NEUTRALITY

The claim that technologies are neutral with respect to values plays a central role in Pitt's thinking about technology.[1] Pitt regards the idea that technologies can embody values of a particular stripe as nonsensical since, in his view, normativity can be located only "in people not in things," and offers three lines of argumentation in support of his view that technologies are neutral (Pitt 2011, 35).

Pitt first argues that technologies are open to a wide range of conceivable uses. This wide flexibility with respect to use in conjunction with normative significance depending on actual consequences supports the conclusion that tools are neutral vis-à-vis particular values, since it should be fairly obvious to everyone that "no matter what the device or system is, if it is put to different uses, with different ends in mind, the consequences of using it will be different" (Pitt 2000, 82).[2] Consider Pitt's thought on the birth control pill, one of the examples he uses to illustrate tool neutrality. Non-neutralists argue that written into the birth control pill are norms with particular burdens and requirements that facilitated separating reproduction from sexual activity and changed perceptions of family planning, events that have given women more control over their bodies and pregnancy, which with other factors have altered the power relationship between men and women. Pitt acknowledges that the birth control pill drove these social transformations but insists that inferring from this that it embodies non-patriarchal norms ignores the fact that the pill was originally developed to help women suffering from irregular menstrual cycles achieve pregnancy. He insists: "[I]f the pill embodies normativity, it is the norms of a tradition in a male dominated society" (Pitt 2011, 35). If nowadays it is true that the pill does not favor patriarchal institutions, this is because people decided to design it for different purposes, not because it embodies values that favor greater sex equality.

Apart from use flexibility and meaning consequentialism constituting evidence of tool neutrality, Pitt is skeptical of approaches to technology that privilege social critique over epistemological issues. These critiques evaluate the alleged consequences of a technology using norms taken for granted as valid, often before adequately understanding the technology being targeted or without offering much in the form of empirical evidence supporting the claims being made about it. However, "understanding what we know about technology, and understanding *how we know that what we know is reliable* are the prerequisites to offering sound evaluation of the effects of technologies and technological innovations on our world and our lives" (Pitt 2000, viii). Without this self-understanding social critiques of technologies create ideological impasses that "[preclude] our ability to resolve whatever disagreement may be at issue, as well as our ability to understand the artifact or system involved" (Pitt 2000, 74). In contrast, recognizing that technologies are neutral with respect to non-epistemic values rightly shifts attention from artifacts to the decision-makers since, on Pitt's view, "it is [their] decisions that form the *loci* of understanding. Technological explanations [should be] explanations of the causes and consequences of decisions made by key players" (Pitt 2000, 65).

Besides these reasons, Pitt sees ascriptions of normativity to artifacts as instances of *ethical colonialism*, a kind of imposition of moral values on others shown by "the attempt to endow everything in the world as an actor with moral value. It is to deny that there are other types of values, which have their own integrity and can operate in an ethically neutral framework" (Pitt 2011, 35). Moral values have a place in the life of technologies since, like epistemic values, they inform the decisions to design or use tools in one way rather than another. But for Pitt this implies only that they are to be found in decision-making processes, not in the artifacts themselves, thus his emphasis on identifying and assessing the decisions of key players in order to understand a technology and its effects. As he puts it, "decision-making is a value-dominated activity. With respect to technological artifacts, whenever they are employed, it is because a decision has been made to do so. Making decisions is an inherently value-laden activity since it always involves making a choice" (Pitt 2011, 35).

This essay endorses Pitt's claim that critiques of technologies should not take for granted their normative framework and offer empirical evidence to substantiate their assessments. It also avoids reifying technology (another Pitt warning) by discussing the philosophical significance of one kind of technology (social networking sites) without favoring any one definition of technology.[3] Pitt's Common-Sense Principle of Rationality (CPR), later discussed in detail, I contend must be part and parcel of any sound approach to the problems technological change inevitably generates.

Pitt's arguments for tool neutrality, however, are not convincing. It is true that the meaning of a technology can be quite flexible, especially in the initial stages of development. But this hermeneutic flexibility often decreases substantially or disappears altogether once initial choices and commitments become fixed in material equipment, economic investment, and social habits. Once particular ways of designing a technology become socially established, this both limits and influences how people view and use it. Returning to the birth control pill, even if it is true that the inventor wanted originally an aid for pregnancy, the objective properties suggested the opposite and, once certain choices and commitments were made, the possibility of functioning like one was foreclosed.[4] Cases like the pill also warn against reducing a technology's significance to the expressed intentions or motives of creators and owners. Not only may a technology end up being something other than was originally intended, but its real or full significance may not become apparent until its psychic and societal consequences become visible and thoughtful people begin identifying and describing them.

Of relevance too is the fact that a technology's visible consequences often expose particular values built into its design that either explicitly or implicitly informed the decisions that created it. Once a technology becomes integrated into important parts of life, it can acquire a momentum of its own, thereby contributing to the reproduction of the particular values it embodies by reinforcing certain patterns of life and valuation. This is a kind of normativity exhibited by technical objects in their own right that must be described adequately to understand the negative consequences being empirically linked to their use, identify the alternative design possibilities that were foreclosed after a particular design became the norm, and how these two relate to the explicit decisions of key players and the full intentions behind these key decisions. It may appear strange to insist that artifacts can have normative properties, not just people or their decisions. But this is because one may be assuming a dualism or discontinuity between mind and matter. A Peircian pragmatist like Pitt, however, should not make this assumption, given that Peirce thought of mind and matter as part of a continuum, often describing matter as mind imbued with indurated habits, which implies that there is no conceptual barrier in claiming that technical artifacts can materialize values (CP 6.277).[5]

The neutrality thesis overlooks the fact that technologies can be disposed to favor particular values by altering habits of mind and behavior in ways that reproduce those values. Acknowledging this truth and letting it inform critiques of technologies need not result in ethical colonialism, which can happen when one "endows everything in the world as an actor with moral value" (Pitt 2011, 34). Ascribing particular norms to a tool and explaining how it can itself propagate them does not involve moral judgments concerning the desirability or undesirability of the tool being this way. Rather, doing so stems from the intuition that technical things have normative properties that are important to identify and describe in order to fully grasp their psychic and societal consequences. If moral judgments are made, for example, that a certain technology must be redesigned because it threatens freedom and democracy, these judgments need not appeal to foreign values—which the term "ethical colonialism" implies—but instead to values shared by the affected communities. Pitt should not be concerned that conceiving artifacts as having normative properties amounts to endowing everything with moral agency. Setting aside current debates about the possibility of strong artificial intelligence and how it might force us to revise the claim that technologies cannot have or develop agency, it is, of course, true that the technologies and systems humanity has invented until now are not entities that can form their own intentions and autonomously act on these intentions. Existing technologies are unable to form value judgments or evaluate the positive and negative changes they introduce into the world. But the kind of normativity this essay is committed to refers to technologies making causal contributions to the reproduction of certain values by means of their impacts on patterns of life and thinking. This conception does not presuppose intentional agencies.

SNS AND THE ETHIC OF PERSONALIZATION

It is doubtful that there is a single definition of "technology," but there are three senses of the term that frequently recur. First, it can denote manufactured objects or physical artifacts such as cups, lamps, and telescopes. Second, the term sometimes refers to the techniques, methodologies, processes, skills, and knowledge required to create and use physical artifacts. Finally, the term may describe the systems that use human labor, physical artifacts, and relevant knowledge and methodologies to manufacture objects, or to accomplish tasks that humans cannot accomplish without the help of these systems.[6] Pitt's characterization of technology as *humanity at work*—a phrase Pitt coined to describe humanity's ongoing effort to create and use tools to meet "ever changing needs and goals" (Pitt 2000, 11)—includes these three conceptions. Apart from the merits of Pitt's phrase, it is clear that major technological innovations in any of these three senses of "technology" entail quantitative and qualitative extensions of our muscular, sensing, mental, and even social functions. Microphones, for example, amplify our voice but airplanes and submarines introduce new possibilities by enabling humans to flight or operate completely submerged under the sea.

It is not hyperbole to claim that the Internet is the defining technology of our time. Now operating at a global scale, this array of computer networks is responsible for many of the societal transformations the world has been experiencing since the Internet's rapid expansion began in the 1990s. Internet's capacity to process and transmit digital information fast, and on demand, has quantitatively and qualitatively altered the way human beings connect, communicate, and process information. It has also reshaped the way human beings socially interact with each other. Due to the many opportunities for connection and communication made possible by the Internet, social interactions have become increasingly organized around personal and organizational networks mediated by social networking sites like Facebook and Twitter (Castells 2014). These technologies are "Internet-based platforms that allow the creation and exchange of user-generated content, usually using either mobile or web-based technologies" (Margetts et al. 2016, 5). SNS empower users to create and maintain relationships virtually by allowing them to specify the list of users with whom they share a connection and view and traverse this list and those made by others within the system (Boyd and Ellison 2007). They provide a range of tools like free personal web pages, news feeds displaying in one place an ongoing list of updates, online groups, text messaging systems, and other applications meant to facilitate connection, sharing, and different kinds of social interactions.

This shift toward social networking online has had many positive effects on individuals and societies. In particular, it has put "people at the center of personal and organizational networks that can supply them with support, sociability, information, and a sense of belonging" (Rainie and Wellman 2012, 124). Many geographically dispersed connections that we have reason to want and value are unimaginable or hard to maintain without the extensions in communication and sociality SNS make possible. Geographically dispersed families can witness each other's lives in detail from a distance. Marginalized or alienated individuals can find community with people living elsewhere. People can join a myriad of support groups existing online, raise awareness about important issues affecting their communities, and initiate social or political movements.

Despite these and other unmentioned benefits, a growing body of literature now indicates this shift also created many problems (Benkler, Faris, and Roberts 2018; Alter 2018; Concheiro 2016; Hassan 2012; Jackson 2008; Pariser 2011; Wolf 2018). The remainder of this section specifies the set of particular values SNS currently embody in order to make evident how their incorporation of these values into their design enables conditions responsible for some of these negative effects; in particular, the empirically documented effects their use has on our capacity for critical self-reflection.

The current design and functioning of SNS embodies an ethic of personalization. What I mean is a set of assumptions about *what ought to matter and why* that takes individualized interactions and contents to be the most valuable, prioritizes the gratification of current desires, and reduces freedom to consumer choice. In other words, a vulgar form of libertarianism implying that the best and most effective way to empower individuals, improve their understanding of the world, their relations to each other, and in general bring a diverse and pluralistic world closer together is with online sites that allow and encourage them to promiscuously generate and share whatever content they see fit (especially detailed data about their private lives and preferences); these sites curate content focusing mostly on what personally matters to users while ignoring the rest, and reinforce these preferences with algorithms that identify and cater to them more effectively and efficiently over time regardless of the social or epistemic value of doing so. The result is "subjective, personal, and unique" news feeds that prioritize private concerns over socially valuable content, widely shared experiences, or possibilities capable of enlarging the perspective of users (Mosseri 2016).

The dominance of personalized SNS reflects the mutually reinforcing interaction between the nature of this type of technology, the pursuit of power and profit within a capitalist system, and the way these two factors reinforce certain human dispositions. By their very nature, SNS like Facebook, Instagram, and Twitter are decentralized, user-centered networking tools meant to make it easy, fast, and cheap for individuals and groups to generate and share information and participate with one another in a wide variety of social activities within a bounded online space. This essential configuration makes users the center of their personal and organizational networks with extensive freedoms to determine what they do and see within these networks: from the quantity and quality of the content they create and share, to how they decide to engage other people's content. We can expect then a certain degree of personalization due to the user-centered nature of SNS. But this is not sufficient to account for the amount of personalization we are now experiencing—like Facebook prioritizing posts from family by putting them on top of our news feeds or algorithms that filter out unengaged content—since one can imagine these tools working in ways that better balance this tendency with diverse news and other content that addresses common interests. The fact that these sites do not currently strike this balance also reflects the pursuit of power and profit within a capitalist system and the way it affects the decisions of key players. In capitalist societies, companies can grow and remain competitive only if they accumulate power and maximize profit. This means that owning and operating a successful social media platform like Facebook, which has over two billion users and is worth billions of dollars, requires growing user numbers, encouraging high volumes of original posts, and collecting large amounts of user data to profit from its micro-advertising value. Personalized SNS are intentionally designed to achieve these goals with news feeds and other functions that give users, often without their explicit knowledge and consent, more and more of what these systems think users want or might want in order to keep them engaged, interested, and encourage them to promiscuously generate and share profitable kinds of information.

Still, this strategy would not be as successful as it currently is in attracting and maintaining high user engagement—a necessary condition of reproducing the logic of personalization—without tacitly appealing to certain facts about human nature; in particular, the human tendencies for *homophily* (to prefer people like us) and *confirmation bias* (to prefer agreeable content). It is well known that human beings prefer to pursue their own interests, are more attracted to people with similar ideas and values, and manifest a stronger tendency to seek confirmation rather than disconfirmation of important beliefs. This is not necessarily problematic. After all, it is difficult to maintain meaningful and fulfilling relationships with people that do not share our interests. In an ever more complex world filled with so much information, it is reasonable to prioritize activities and interests that we identify with. Cognitive dissonance is painful and takes a lot of energy to transcend, so it is hardly surprising that humans prefer ideas and voices that conform to their existing beliefs. Given these facts about human nature, it is desirable and, to some extent, unavoidable for owning companies to personalize news feeds independently of the goal of maximizing profit. But when these companies indulge these habits to maximize power and profit without balancing it with broader, more diverse content that enlarges experience, invites self-criticism, and informs on social issues, the result is likely to be highly personalized platforms whose workings generate conditions that reproduce the negative impacts of personalization at the expense of important social ends.

One of these important social ends—at least within societies that claim or aspire to be democratic—is nurturing a technological environment that encourages people to engage critically with their beliefs, both as a condition of genuine autonomy and well-functioning democratic institutions. Personalized SNS threaten the aim of nurturing this kind of technological environment by reproducing conditions that compromise a necessary feature of critical self-reflection; more precisely, responsiveness to reasons (what I describe later as being moderately responsive to reasons, or MRR). Understanding how personalized SNS threaten the ability to moderately respond to reasons is a prerequisite to press for a technological milieu more compatible with conditions that foster genuine autonomy and well-functioning democracies, the latter implying an informed citizenry able to identify the real issues facing their society and the species, choose adequate ways to address them, and elect people willing and able to translate these solutions into law. Pitt insists correctly that claims about the impact a given technology has on self and society must be supported by empirical evidence. He notes: "[I]f we don't know what kind of knowledge is being invoked, then it is hard to assess the particular criticism. If we can't assess the criticism, then it is unclear how to incorporate it into our thinking about the way the world works and what actions we should take" (Pitt 2000, 71). The next section builds on this important demand and describes empirical studies indicating that the threat personalized SNS pose to critical self-reflection, hence to the end of fostering genuine autonomy and well-functioning democracies, is real and must be addressed.

SNS AND EPISTEMIC BUBBLES

Recent findings indicate that personalized SNS sustain environments that are antithetical to critical self-reflection, such as epistemic bubbles. Epistemic bubbles are informational structures that exclude alternative voices and sources of information. They are insular, self-contained, homogenous clusters of like-minded people that are, for the most part, committed to similar beliefs and points of view (Nguyen 2018; Sunstein 2017; Jamieson and Capella 2008). Online they take the form of personal, unique, subjective news feeds that mostly cater to individual preferences, beliefs, and preconceptions while simultaneously filtering out alternative views and sources of information. Whether the news feed of a Facebook or Twitter user fits this description partly depends on their personal interests and preferences, those of their friends and followers, and other types of connection. Someone who has broad interests, a diverse number of friends, and actively seeks exposure to challenging ideas, is unlikely to be living in an epistemic bubble. But the reality is that many of us prefer to talk to people who think like us, read information that agrees with our beliefs, and share personally significant content regardless of its broader educational or social value. When these human proclivities for homophily and confirmation bias interact with the unprecedented control we have over our personal networks, the easiness to which we can ignore online connections that we dislike, algorithms that prioritize "meaningful" engagements in order to maximize profitable user information, and other factors that subtly homogenize our news feeds, the most likely result is that many of us intentionally or inadvertently will end up being trapped inside epistemic bubbles.

This observation is borne by empirical studies on epistemic bubbles in Facebook and Twitter. In 2015 Bakshy, Messing, and Adamic published in *Science* a significant study on political polarization among Facebook users. They wanted to know how online networks influence exposure to perspectives that cut across ideological lines. The researchers examined how 10.1 million U.S. Facebook users interacted with socially shared news. They measured ideological *homophily*—our urge to cavort with those similar to ourselves—by asking users to declare their political affiliations. Additionally, the study examined the extent to which heterogeneous friends could potentially expose users to crosscutting contents and cleanly separated suppression of crosscutting contents attributable to the Facebook algorithm from suppression attributable to a user's clicking behavior.

The study found that the Facebook algorithm suppressed exposure to crosscutting content by 8 percent for self-identified liberals and 5 percent for self-identified conservatives. These are not large numbers, but they do show that Facebook users were seeing fewer oppositional news items solely because of the effects of the algorithm; and this is before Facebook's decision to prioritize posts from family and friends. Moreover, when individual behavior was counted, filtering increased an additional 6 percent for liberals and 17 percent for conservatives. The authors' main conclusion was that "compared with algorithmic ranking, individual choices played a stronger role in limiting exposure to crosscutting content" (Bakshy, Messing, and Adamic 2015, 1132). Nevertheless, when it is considered that the study found clear evidence of confirmation bias and that only 7 percent of the content users click on is actually hard news; the most important lesson should rather be that Facebook and relevant human tendencies, like *homophily*, work synergistically to narrow the field of vision of users.

Another study on Facebook led by Michela Del Vicario and her coauthors (2015) tested whether users engage in the kind of self-sorting that generates epistemic bubbles. They measured how Facebook users spread conspiracy theories (using thirty-two public web pages), interacted with science news (using thirty-five such pages) and "trolls," automatic accounts that intentionally spread disinformation (using two webpages). After analyzing everything posted on Facebook by research subjects during a five-year period, the researchers found virtual equivalents of gated communities. They discovered that conspiracy theories and disinformation tend to spread rapidly within these clusters. Interestingly, when users were exposed to clearly unrealistic and satirical claims, for example, the claim that chemtrails contain *Sildenafil Citratum* (the active ingredient in Viagra™), many of them liked and shared the claims agreeable to their beliefs. In fact, people "mostly selected and shared content according to a specific narrative and ignored the rest" (Del Vicario 2015, 4). The main conclusion drawn from this study is that "selective exposure to content is the primary driver of content diffusion and generates the formation of homogeneous clusters" (Del Vicario 2015, 4).

In a different study Conover and his coauthors (Conover et al. 2011) also discovered politically polarized bubbles on Twitter after they analyzed "retweet" and "mentioned" networks ("mentions" are tweets that contain another Twitter user's @username), comprising more than 250,000 tweets from more than 45,000 users, during the six weeks leading up to the 2010 U.S. election. The main finding was a high degree of polarization in the network of retweets with hardly any interaction between "left- and right-leaning users" (Conover et al. 2011, 89). Because the network of mentions had the structure of "a single politically heterogeneous cluster of users in which ideologically-opposed individuals interacted at a much higher rate compared to the network of retweets," polarization in the mentioned network was less severe. Despite this difference, the authors conclude that "political segregation, as manifested in the topology of the retweet network, persists in spite of substantial cross-ideological interaction" (Conover et al. 2011, 89).

The key lesson of these studies is that personalized SNS like Facebook and Twitter facilitate the formation of epistemic bubbles and in this way reproduce the values of personalization they embody by magnifying and often deepening the adverse effects epistemic bubbles can have on individuals and societies. One of these negative effects is *polarization*, which can alienate individuals from each other and also entrench existing societal divisions. Polarization intensifies opinions and attitudes so that they are held more strongly (for or against a given issue) in one group than in another. It increases the distance between people on an issue, usually because one or both of the parties involved come to adopt an extreme version of their initial views. If my position on immigration becomes more extreme, but yours remain the same, then our stances on immigration have become more polarized. The distance between our beliefs on immigration has increased.

Research indicates that when people discuss ideas only with like-minded others, they often end up endorsing extreme versions of their initial beliefs. Consider a study involving ten face-to-face groups, each composed of six like-minded people (five conservative and five liberal), that were asked by Schkade, Sunstein, and Hastie (2007) to deliberate on three controversial issues: the right of same-sex couples to enter civil unions, affirmative action, and whether the United States had an obligation to sign international treaties intended to combat global warming. The anonymous opinion of each individual was recorded on these three issues before and after fifteen minutes of group discussion, and each group was asked to reach a public verdict prior to individual members of each group making their final anonymous statements. The study found that members of each group adopted extreme versions of their initial beliefs after discussion. Civil unions, for example, became more popular with liberals but decreased in popularity among conservatives. After discussion, the groups themselves become more internally homogenous, which increased polarization between them (Schkade, Sunstein, and Hastie 2007, 917). Sunstein notes that "before discussion, some liberal groups were, on some issues, fairly close to some conservative groups. The result of discussion was to divide them far more sharply" (Sunstein 2017, 69).

EPISTEMIC BUBBLES AND RESPONSIVENESS TO REASONS

A hitherto overlooked but important way the patterns observed in these studies replicate the ethic of personalization is by disrupting people's capacity to respond to reasons. In a different context, Fischer and Ravizza (1998) argue that the ability to make morally responsible choices—for an entity to be an apt target of praise, blame, and other "reactive" attitudes—presupposes *guidance control*—the power "to be in charge" of the thinking that led to a decision.[7] Infants, nonhuman animals, severely mentally impaired individuals, and robots in their current state of development cannot be held morally responsible for what they do or fail to do (they cannot be considered apt targets for praise, blame, resentment, indignation, and other reactive attitudes) because they don't have guidance control of their thinking and behavior. What is required for someone to have guidance control? What are conditions necessary and sufficient for having the capacity to be in charge of one's decisions, or the thinking that led to a decision? On Fischer and Ravizza's model of moral responsibility, an agent has guidance control of decisions when her decision-making faculty is "moderately reasons-responsive" MRR (Fischer and Ravizza 1998, 69). A decision-making faculty (someone's mind, if you will) is MRR if it is regularly receptive to reasons—where "regular" means the decision-making faculty manifests a "minimally comprehensible pattern of recognition" of reasons—and is at least weakly reactive to recognized reasons, meaning that it shows the ability to act on recognized practical reasons in at least one counterfactually hypothesized case pinpointing practical reasons sufficient for acting other than one did (Fischer and Ravizza 1998, 69–76).[8] A creature whose decision-making faculty does not show a minimally comprehensive pattern of reasons recognition—that fails to regularly register relevant reasons for alternative courses of action, or fails to rank-order them consistently—or is unable to choose and act on recognized reasons for doing otherwise in at least one counterfactually hypothesized case—cannot be considered to be a morally responsible agent.

Scholars disagree about the adequacy of MRR as a model of moral responsibility, in particular, its requirement of weak reactivity. Weak reactivity only requires agents to be able to do otherwise in at least one case pinpointing sufficient reasons to do so. But, as Mele (2006) and others have noted (Schoonover and Guajardo 2018), a person may manifest this capacity yet be nearly always—except in uniquely extreme situations—incapable of converting reasons into actions. Consider the case of the agoraphobic man discussed by Mele (2006, 150–52). The challenge this man faces isn't that he doesn't regularly recognize valid reasons for leaving the house, or that he can't rank-order them consistently. It is rather that he can react only to reasons evoked by extreme circumstances, for example, discovering that his house is on fire or dangerously floating during a storm surge. Yet, someone handicapped to this degree meets the threshold of weak reactivity since it only requires being able to do otherwise than one did in at least one case, which by hypothesis Mele's agoraphobic can do when faced with emergencies. But this counterintuitively implies that one can hold someone like this morally responsible for not attending their sister's wedding, even if it is obvious that such a person is not capable of reacting to reasons for doing so. The same holds for moral responsibility shortfalls among people with autism (Stout 2016). In such cases, individuals are weakly reactive but incapacitated in ways that may preclude treating them as morally responsible agents.

Despite this challenge, which I believe can be met by a stronger reactivity requirement, Fisher and Ravizza are right: guidance control of one's thinking and its outputs (beliefs, decisions, and so on) requires regular recognition of reasons, the ability to rank-order them consistently and to translate at least some of these recognized reasons into action. If guidance control of one's thinking and actions presupposes MRR, and critical self-reflection is not possible without guidance control, it follows that technologically induced disruptions of MRR compromise critical self-reflection, thereby hindering efforts to design technological milieus compatible with genuine autonomy and democracy. Let me explain some of these disruptions before I draw connections to Pitt's thoughts on technology.

Epistemic bubbles can work against MRR in important ways. Since they exclude crosscutting content, they keep people from recognizing potentially relevant reasons they could have recognized had these reasons appeared before their mind. John Stuart Mill expressed this simple but important truth well when he noted that "facts and arguments, to produce any effects on the mind, must be brought before it" (Mill 1972, 79). Epistemic bubbles also reinforce confirmation bias. People driven by confirmation bias pay attention to content that confirms their beliefs and ignore conflicting facts and opinions. This increases the chances that people so disposed will fail to recognize countervailing reasons. But even if people so disposed were to grasp countervailing reasons they may still *partially* rank-order them, for example, by giving more credence to agreeable facts and ideas. Epistemic bubbles can also be purveyors of disinformation (information that is intentionally false or misleading), like a piece of fake news or a decontextualized photo. It is unclear how much influence fake news, conspiracy theories, and misleading information can have on those who see them. Some studies indicate that most people ignore the most blatant instances that appear on their news feeds, with the exception of senior citizens and ultra conservatives (Guess, Nadler, and Tucker 2019). But disinformation can be subtle: a decontextualized photograph, a skillfully doctored video, or a comment from what appears to be a genuine news outlet. Hence epistemic bubbles also increase the chances of someone mistaking false or misleading data for a reason; they can induce *misrecognition* of reasons.

So far it has been claimed that the news feeds that work as epistemic bubbles can prevent users from exercising guidance control of their beliefs by excluding relevant reasons, disposing them to ignore or partially rank-order them, or by inducing misrecognition of reasons. None of these three possibilities, however, entails a complete loss of guidance control. Were this to occur, it would signal a total loss of capacity for moral responsibility. In other words, a person trapped in an epistemic bubble may be able to recognize excluded reasons had they been exposed to them, rank-order them impartially in other contexts, and also have the ability to convert them into action. Nevertheless, it is possible that psychic effects like polarization could undermine someone's capacity to recognize, rank-order them impartially, or react to reasons that *challenge dogmatically or fanatically held beliefs*. Studies on polarization, like the one I described earlier, point to overconfidence as the major cause of polarization. Polarization occurs when someone adopts an extreme version of a belief after becoming more confident even if it happens in response to invalid reasons, such as finding that others think like them, or by being disproportionately exposed to more arguments in favor of their beliefs than arguments against them. Overconfidence not only discourages critical self-reflection, which prevents people from recognizing reasons they might have otherwise recognized, but can also result in dogmatism and fanaticism. Ardently or inflexibly held beliefs can undermine people's capacity to respond to reasons against them by rendering those who hold them in this way unable to recognize, or impartially weight, or react to these reasons. This forecloses the possibility of changing these beliefs when warranted, or the extreme inflexible mode in which they are being held.

These possible ways of disrupting MRR reproduce the ethic of personalization by trapping people in epistemic bubbles that discourage the kind of self-reflection that is required to question the personal and social value of being mostly exposed to costume-tailored, agreeable content. It is a fact that critical self-reflection is rarer than the view of humans as rational agents assumes. This is not just due to people being predictively irrational, as Ariely (2008) and others argue, but also because critical thinking takes energy and the human brain likes to save energy by acting on habits. Still, it does not follow that artifacts should promote this tendency, yet SNS do so by feeding habits that, in turn, directly inhibit any disposition for critical self-reflection or the factors that encourage it.

PITT'S COMMON SENSE APPROACH TO TECHNOLOGICAL PROBLEMS

Without redesigning SNS in ways compatible with MRR, we can expect this problem to continue. It is time to discuss how some relevant ideas in Pitt's philosophy of technology may help in dealing with technological problems like this one. Central to Pitt's approach to technological problems is the idea that the process of design must be a communal, self-correcting process capable of effectively responding to experience.[9] He argues that technological change inevitably produces bad unintended consequences, and it is not possible to anticipate them all or find ways to fully ensure prior to acting that the decision to adopt new technologies, or redesign existing ones, will lead to success without risking bad consequences. We must learn how to live with technologies so the key to rational technological choices—choices that allow us to harness the possibilities opened up by a new technology while effectively dealing with its negative effects—is not a demand for methods that guarantee success (they do not exist) but to follow the Common-Sense Principle of Rationality, or CPR, by building into technological design a feedback loop mechanism that enables designers and users of a given technology to effectively learn from their mistakes. As he says, "to avoid repeating mistakes, i.e., to be rational requires that we learn from those mistakes, that we update our information, eliminating incorrect data and assumptions that led to failure, and then try again, reevaluating those results, and continue in that fashion" (Pitt 2000, 23).

Taking CPR—the injunction to learn from experience—seriously as the guiding principle in humanity's quest to learn to live rationally with technologies by making better personal and collective technological choices also requires teaching people how to apply the design process to their own choice of tools (Pitt 2011, 5–10). After all, if the design process must be a communal, self-correcting process capable of effectively learning from experience, we should expect common sense not only from the creators and owners of tools we must also expect it from users. This means that people must develop some sense of how experience-sensitive design works, how it can improve their own choice of what tools to incorporate into their lives, and how this understanding can function as a framework for demanding better design decisions from the creators and owners of technologies.

These are important recommendations. Pitt (2011) correctly argues that to a large extent human beings are "technological artifacts" (Pitt 2011, 6–7). What individuals and collectives are or may become is largely a function of the technologies they incorporate into their lives. To use Pitt's own words, "in choosing to employ certain artifacts and technologies in your life they make you who you are" (Pitt 2011, 7). Given that the character of our individual and collective lives depends on the quality of our technological choices, teaching individuals and collectives what experience-sensitive design looks like, and how to apply the process to the choice and evaluation of tools becomes extremely important. Such teaching involves giving people tools to clarify ends and values, generate goals, break down those goals into manageable parts, and gauge progress in meeting those goals by means of feedback loop mechanisms built into the design process in order to learn from mistakes. It aims to encourage active participation in the technological design of our lives and the technological milieus we are part of ,i.e., to encourage people to intentionally and reflectively "attempt to design the person they want to become in much the same way we design an artifact" (Pitt 2011, 3). In sum, to equip us with knowledge and tools so we can learn how to be *autonomous self-designers*, that is, beings capable of choosing technological enhancements in accordance with freely formed preferences and values.

Pitt's overall approach to technological problems, then, emphasizes design processes and teaching practices aiming at improving individual and collective technological decisions over time with knowledge, tools, and feedback loop mechanisms that incorporate CPR, or the kind of pragmatic rationality that CPR presupposes: a notion of rationality that stresses learning from mistakes.[10] This approach gets the general goal right—learning how to live rationally with technologies—and identifies some essential means to do so, yet it requires supplementation.

Pitt's approach, for one, must take into account that technologies can propagate in their own right the specific values embodied in their design by means of their psychological and societal impacts. This is especially important when dealing with technologies like personalized SNS which hinder people's capacity to learn from experience by disrupting necessary conditions for critical self-reflection. After all, it stands to reason that people cannot be expected to make more rational technological choices, or demand them from creators and owners, if technologies that have come to be indispensable to function, let alone thrive, in contemporary life—as is the case with SNS—work against CPR by disrupting in subtle ways their capacity to respond to reasons. In practice, taking into account that technologies can and often do embody specific values with profound effects on individuals and societies means that the communal, self-correcting, experience-sensitive design process Pitt recommends as the way to deal with technological problems must foster awareness of the specific values a given technology might come to embody, and how they may serve or fail to serve broader ends and values the community takes itself to be committed to. For this to happen, a collective effort is needed to teach designers and owners of technologies (who often think technological innovation is intrinsically desirable, or a value-neutral endeavor solely responsive to "real" problems) as well as users, ways to recognize and evaluate the often subtle and invisible effects a technology's normative properties can have on individuals and societies.

Pitt's approach must also be placed within a larger critique of how the political economies within which technologies exist relate to the specific values they embody, and what this relation tells us about what it would take to solve the problems empirical research indicates they are causing. As I claimed before, the ethic of personalization SNS have come to embody reflect their nature, human tendencies like homophily and confirmation bias, and the struggles for power and profit within a capitalist political economy. Personalization makes money, and the companies that own SNS, like Facebook, use that money to monopolize their market and capture the political process to protect their interests. Here the issue is not just that the design process being used to develop and operate these sites may be insufficiently responsive to experience, but also that these companies are unwilling to adopt the measures needed to solve the problems these tools are currently creating. Doing so may require them to relinquish the amount of power they have accumulated and adopt something like the more communal, self-correcting design process Pitt's approach favors. But this seems to amount to a recognition that having design processes and teaching practices that make designers and users aware of mistakes is not sufficient to adequately address them, since doing so may require adopting practices and solutions that are not possible without radical systemic changes.

CONCLUSION

Technologies are not neutral. Although there is no way to show a priori that their normatively significant effects on individuals and societies are inevitable or unavoidable, technologies can nevertheless acquire a momentum of their own that facilitates the reproduction of the particular values they embody by, among other ways, reinforcing certain behaviors and thinking habits. SNS are a case in point. These tools encapsulate an ethic of personalization being replicated by epistemic bubbles that disrupt MRR, a necessary condition of critical self-reflection. Pitt's common-sense approach to technological problems must take this into account and be part of broader analyses linking the specific values dominant technologies can materialize in the larger political economy. Let me end this essay by relating these claims to some proposed ways to redesign personalized SNS in order to deal with their adverse effects, like the impediments to critical self-reflection described here.

The ethics of personalization embodied by SNS reflects people's increasing power to filter what they see, and also providers' growing power to filter for each of us, based on what they know or think they know about us. This essay endeavored to explain how these two powers work synergistically to create epistemic bubbles whose adverse psychic effects, like polarization, can impede critical self-reflection by hindering—at least in the ways I specified earlier—people's capacity to moderately respond to reasons (MRR). As far as concrete solutions to this particular problem go, I am convinced we must follow Cass R. Sunstein's (2017) advice: first, providers must design their algorithms to include the goal of exposing users to materials they would not have necessarily chosen or sought themselves. I have in mind unchosen, unanticipated, *serendipitous* ideas and other materials that might nevertheless "change their lives in the right way" (Sunstein 2017, 7). The idea is simple but powerful: people cannot be expected to increase their awareness and hopefully grow intellectually from it unless their news feeds expose them to ideas and sources of information that can loosen their habitual ways of perceiving and thinking, exposing them to experiences and ideas generative of new doubts and questions. But this is not sufficient: people from different backgrounds and identities must be able to communicate, identify common goals, and agree on the relevant facts and methods for solving common problems. This is not possible if there are not common experiences to appeal to, or enough of them to facilitate cross-cultural communication. Algorithms and other functions and apps must play their part in generating at least some of these common experiences for all of us.

SNS are here to stay. The challenge is to learn to live rationally with them, part of which is to balance personalization with broader social ends in the ways just specified. But the causes of personalization are not just technical. The particular problems associated with technologies that embody this ethic point to the difficult problem of how to make "the community of actors whose role is to decide the general policy that in circumstance *C* it is best to do *X*" more socially responsible, and users more intentional and reflective about their choice of artifacts (Pitt 2000, 20). Pitt's common sense approach to technological problems points to part of the solution: design processes and ways of thinking about individual and collective technological choices capable of guiding humanity in efforts to make progress in learning how to live rationally with technologies. It is an open question, however, as to whether or not this way of thinking about technology can become a practice within the confines of a system that tends to prioritize profit and power over social ends.

NOTES

1. Pitt distinguishes epistemic from non-epistemic values. Epistemic values like objectivity, measurement, and justification guide humanity's search for knowledge, while non-epistemic values (aesthetic and moral values in particular) guide the search for the good life. Pitt calls values guiding the search for the good life "aesthetic values rather than ethical or moral values because I see ethical values as ultimately deriving from broader aesthetic considerations" (Pitt 2011, 14). Pitt also considers epistemic values objective, and neutral with respect to aesthetic and moral values. Thus, when he asserts that technologies are value neutral, he means that they are neutral with respect to aesthetic or moral values.

2. Pitt more recently makes this point as follows: "the normative significance is a direct function of how people choose to view them and use them. It is the use to which artifacts are put that exhibits the normativity of the users, not the things" (Pitt 2011, 35).

3. Pitt argues that discussions of "Technology" per se, rather than specific technologies, treat an otherwise complex and pluralistic phenomena as a monolithic thing endowed with a force of its own. Among other problems, this approach "moves the discussion, and hence any hope of philosophical progress, down blind alleys" (Pitt 2000, 87). This essay agrees.

4. To offer two more examples, the bicycle might have acquired a different form early in its development but this initial latitude with respect to its identity has decreased substantially after all the possibilities that were alive at the time gave way to the few that became the norm (Goliski 2005). Similarly, the freedom to redesign SNS to give them different functions from the ones they now serve is likely to be

much less than early in their development, given that certain choices are now built into their current design, money has been invested, and users have become habituated to it.

5. The idea that mind and matter are continuous is an essential component of Peirce's synechism, which "insists upon the idea that continuity [is] of prime importance in philosophy and, in particular, upon the necessity of hypotheses involving true continuity" (CP 6.169). Accordingly, claiming that normativity can be located only in people, not in things, implies some form of dualism between thing and value, but pragmatism is supposed to help thinking transcend dualisms of any kind.

6. Kline (2003) calls systems that produce objects "Sociotechnical Systems of Manufacture" and systems that enable human beings to accomplishing tasks they cannot perform unaided "Sociotechnical Systems of Use" (Kline 2003, 210–11).

7. Moral responsibility here refers to capacities necessary and sufficient for being a morally responsible agent at all—and therefore for being a proper target of praise and blame rather than to judgments about the morality of any action or decision.

8. Fischer and Ravizza (1998) operationalize reasons-reactivity as a counterfactual possibility to act other than one did had there been practical reasons to do so.

9. To quote: "[T]he question of the rationality of the choice of what to do is no longer directed to the individual so much as to the community of actors whose role is to decide the general policy that in circumstance *C* it is best to do *X*" (Pitt 2000, 20).

10. Pitt stresses that "a nice feature of CPR is that it separates being rational from being successful" (Pitt 2000, 22).

REFERENCES

Alter, Adam. 2017. *Irresistible: The Rise of Addictive Technology and the Business of Keeping Us Hooked*. New York: Penguin Books.

Ariely, Dan. 2008. *Predictably Irrational: The Hidden Forces that Shape Our Decisions*. New York: Harper Collins Publishers.

Bakshy, Eytan, Solomon Messing, and Lada Adamic. 2015. "Exposure to Ideologically Diverse News and Opinion on Facebook." *Science* 348 (6238): 1030–32.

Benkler, Yochai, Robert Faris, and Hal Roberts. 2018. *Network Propaganda: Manipulation, Disinformation, and Radicalization in American Politics*. Oxford: Oxford University Press.

Boyd, Danah M., and Nicole B. Ellison. 2007. "Social Network Sites: Definition, History, and Scholarship." *Journal of Computer-Mediated Communication* 13 (1): 210–30.

Bunge, Mario. 2003. "Philosophical Inputs and Outputs of Technology." In *Philosophy of Technology: The Technological Condition: An Anthology*, edited by Robert C. Scharff and Val Dusek, 172–82. Oxford: Blackwell Publishing.

Castells, Manuel. 2014. "The Impact of Technology on Society: A Global Perspective." *Ch@nge: 19 Key Essays on how The Internet is Changing Our Lives*. OpenMind. Adobe PDF eBook.

Concheiro, Luciano. 2016. *Contra El Tiempo: Filosofía Practica Del Instante*. Barcelona: Anagrama.

Conover, M. D., Jacob Ratkiewicz, Matthew Francisco, Bruno Goncalves, Filippo Menczer, and Alessandro Flammin. 2011. "Political Polarization on Twitter." *Proceedings of the Fifth International Association for the Advancement of Artificial Intelligence: Conference on Weblogs and Social Media*: 89–96. https://www.aaai.org/ocs/index.php/ICWSM/ICWM11/paper/viewFile/2847/3275.

Del Vicario, Michela, Alessandro Bessi, Fabiana Zollo, Fabio Petroni, Antonio Scala, Guido Caldarelli, H. Eugene Stanley, and Walter Quattrociocchi. 2015. "Echo Chambers in the Age of Misinformation" (Unpublished manuscript, December 22, 2015). https://arxiv.org/pdf/1509.00189.pdf.

Ellul, Jacques. 2003. "The 'Autonomy' of the Technological Phenomenon." In *Philosophy of Technology: The Technological Condition: An Anthology*, edited by Robert C. Scharff and Val Dusek, 386–98. Oxford: Blackwell Publishing.

Feenberg, Andrew. 1999. *Questioning Technology*. London and New York: Routledge.

Fischer, John Martin, and Mark Ravizza, S. J. 1998. *Responsibility and Control: A Theory of Moral Responsibility*. Cambridge: Cambridge University Press.

Golinski, Jan. 2005. *Making Natural Knowledge: Constructivism and the History of Science*. Chicago: The University of Chicago Press.

Guess, Andrew, Jonathan Nagler, and Joshua Tucker. 2019. "Less Than You Think: Prevalence and Predictions of Fake News Dissemination on Facebook." *Science Advances* 5 (1). DOI: 10.1126/sciadv.aau4586.

Hassan, Robert. 2012. *The Age of Distraction: Reading, Writing, and Politics in a High-Speed Networked Economy*. New Brunswick, New Jersey: Transaction Publishers.

Jackson, Maggie. 2008. *Distracted: The Erosion of Attention and the Coming of the Dark Age*. Amherst, New York: Prometheus Books.

Jamieson, Kathleen Hall and Joseph N. Cappella. 2008. *Echo Chamber: Rush Limbaugh and the Conservative Media Establishment*. New York: Oxford University Press.

Kline, Stephen J. 2003. "What is Technology?" In *Philosophy of Technology: The Technological Condition: An Anthology*, edited by Robert C. Scharff and Val Dusek, 210–13. Oxford: Blackwell Publishing.

Margetts, Helen, Peter John, Scott Halem, and Taha Yasseri. 2016. *Political Turbulence: How Social Media Shape Collective Action*. Princeton, NJ: Princeton University Press.

McLuhan, Marshall. 2001. *Understanding Media: The Extensions of Man*. London and New York: Routledge.

Mele, Alfred. 2006. *Free Will and Luck*. Oxford: Oxford University Press.

Mill, J. S. 1988. *Utilitarianism, On Liberty, and Considerations on Representative Government*. London: Everyman's Library Classics.

Mosseri, Adam. "Building a Better News Feed for You." *Facebook Newsroom*, June 29, 2016. https://newsroom.fb.com/news/2016/06/building-a-better-news-feed-for-you/.

Nguyen, C. Thi. "Escape the Echo Chamber." Published by *Aeon* on April 9, 2018. https://aeon.co/essays/why-its-ashard-to-escape-an-echo-chamber-as-it-is-tofee-a-cult.

Pariser, Eli. 2011. *The Filter Bubble: What the Internet Is Hiding from You*. New York: The Penguin Press.

Peirce, Charles S. 1931–1958. *The Collected Papers of Charles Sanders Pierce*. 8 vols. Edited by C. Hartshorne and P. Weiss (Volumes 1–6) and A. Burks (Volumes 7–8). Cambridge, MA: Harvard University Press.

Pitt, Joseph C. 2000. *Thinking About Technology: Foundations of the Philosophy of Technology*. New York and London: Seven Bridges Press.

Pitt, Joseph C. 2011. *Doing Philosophy of Technology: Essays in a Pragmatic Spirit*. London and New York: Springer.

Rainie, Lee, and Barry Wellman. 2012. *Networked: The New Social Operating System*. Cambridge, MA: The MIT Press.

Schkade, David, Cass R. Sunstein, and Reid Hastie. 2007. "What Happened on Deliberation Day." *California Law Review* 95 (3): 915–40. http://dx.doi.org/https://doi.org/10.15779/Z38740Z.

Schoonover, S. B., and Ivan Guajardo. 2019. "Why C-luck *Really* Is a Problem for Compatibilism." *Canadian Journal of Philosophy* 49 (1): 48–69.

Stout, Nathan. 2016. "Reasons-Responsiveness and Moral Responsibility: The Case of Autism." *Journal of Ethics* 20 (4): 401–18.

Sunstein, Cass R. 2017. *# Republic: Divided Democracy in the Age of Social Media*. Princeton, NJ: Princeton University Press.

Vaidhyanathan, Siva. 2018. *Anti-Social Media: How Facebook Disconnects Us and Undermines Democracy*. Oxford: Oxford University Press.

Winner, Langdon. 2000. "Do Artifacts Have Politics?" In *Technology and the Future*, edited by Albert H. Teich, 150–67. Boston and New York: Bedford.

Wolf, Maryanne. 2018. *Reader, Come Home: The Reading Brain in the Digital World*. New York: Harper.

Chapter 7

A Celtic Knot, from Strands of Pragmatic Philosophy

Thomas W. Staley

I've been asked to provide some reflections about my friend and adviser, the philosopher Joe Pitt.[1] I'm sure that other participants in this collection are far more capable of unpacking Pitt's many conceptual contributions to the field, especially in the areas of technology and science. They will tell you about Sicilian realism, and humanity at work, and much more, in better form than I ever could. Instead of trying to follow that path myself, I would like to talk more broadly about pragmatism. In particular, I would like to think through Pitt's version of pragmatism pragmatically, by looking at the practice of philosophy as I've experienced it through my own relationship to Pitt's work—by which I mean his daily efforts, rather than his publications. That perspective means that I will necessarily be telling a partial version of the story that I think matters here. Each of Pitt's other former students could probably relate their own parallel narrative along these lines, and in summing those up I think that we might collectively approach an accurate picture of what "the philosophy of Joe Pitt" amounts to at the end of the day—not so much as a matter of intention or conception but as a matter of practice.

I'll try to begin at the beginning. In the late 1990s, as part of the doctoral program in Science and Technology Studies at Virginia Tech, I participated in a series of three courses in philosophy led by Pitt. Each focused on a single text: Peter Galison's *Image and Logic: A Material Culture of Microphysics*—at the time just published in 1997; David Hume's classic *A Treatise of Human Nature* (1738–40/1978); and Wilfrid Sellars's (1963) *Science, Perception and Reality*, which collects eleven essays written by that scholar. Structurally, each of these semester-long experiences was remarkably similar: a close, slow reading of these works in their entirety, start to finish, with both weekly response writing (to be hacked apart by the master and returned) and subsequent talk around a table, with participating students rotating through "leading"—perhaps more accurately "priming"—that discussion (with Dr. Pitt at the head of the table). This was our practice. This, at that moment, was what philosophy was like. Or where it began.

There are few obvious similarities among the three books just mentioned. I can't exactly recall the order in which we read them (likely as listed above, I believe), but that may be irrelevant. Each of them was given the same treatment by those of us involved: We tried to tackle them, "no holds barred," which is to say that we were encouraged to wrestle with our own relationship to the words and ideas as directly as we could, and to find or develop a sense of their relevance—to us, to others, in context. Within this microscopic attention (Pitt being an enthusiast of microscopy, see Pitt 2011a, 2019) to the details on the page, we hoped to find pieces of some bigger thing(s). Each episode had the potential to provide some evidence of "how things in the broadest possible sense of the term hang together in the broadest sense of the term" (Sellars 1963, 1). Each of us brought, and took away, different things to and from the table—it was a collective mental feast or festival in that regard (Pitt being an enthusiast of feasts and festivals). It was also work.

Beginning with that experience, I'd like now to reflect on some of the ways these texts and that way of dealing with them have shaped my own trajectory over the past two decades. Part of this will involve exploring how Galison, Hume, and Sellars might—in fact—"hang together" to some extent (or how they feed back upon one another perhaps). Another part, relatedly, will aim at some lessons in the practice of philosophy as I've learned it from Pitt. Please bear with me as I recall a series of events that may not necessarily have been exactly sequential. That mode of storytelling will show up again in other hands later on.

But first, Galison. *Image and Logic* is a lengthy and detailed treatment of twentieth-century research traditions in particle physics; a rigorous—if slightly romanticized—history of how that work was done, how it was talked about, and how it was transmitted across and among different groups. It weaves multiple interpretive strands together, but among other things, it emphasizes: (1) a dichotomy between viewing experimental phenomena (the "image" pole) and systemizing theoretical interpretations of them (the "logic" pole); (2) the intrinsic mediating role of language in knitting together those aspects of the science; and (3) the diversity and locality of that linguistic process—something that Galison describes through the notion of "creolization," or hybridization of words and ideas as they move from one circumscribed research community to another.

For someone like myself, having recently come from laboratory work much like that being described on the page, these concepts were at the same time strikingly familiar and quite revelatory. Familiar in the sense of practice (I had done just such things), but also revealing in their formulation (I had never thought about them quite so). Pitt and I began to talk about this, and more work ensued. My own central impulse was to use these insights to probe further the relationships between sensory experience and scientific theorizing. Given Pitt's own intellectual investment in the role of visual instruments (microscopy again, as well as telescopes, see Pitt 2011b) in the history of technoscience, this provided us with a sort of common language, or at least the prospects of our own "creole" or communicative jargon. This was a gateway to the domain of sensory technologies.

Many people will tell you that science is a fundamentally visual enterprise: As a practical matter, seeing is believing, and the realities of the world must be literally shown to be confirmed. There is, of course, something to this. Herein lies the sense of Galileo's demonstration of the Jovian moons, which Pitt has amply discussed in print as an inception point for technoscience (Pitt 1992). Galison's framing of this same kind of process through the image-logic dipole neatly encapsulates the notion: We can talk about things all day in terms of a logical framework, but failing an experiential access point—the technologized eye of the human, assisted as necessary by instruments and mechanisms—there's a disconnect from "reality." Good as far as that goes in centralizing technology as part of knowledge-making, and humanizing that process. But the plot thickens.

One of my own connections to such issues, prior to wrestling with Pitt and Galison on the matter, involves the question of how to find patterns in sets of data. This isn't just something scientists do every day; it's something everyone does every day: People are pattern-making creatures, whether the patterns in question apply to a list of numbers or not. But when we do get down to numbers and logic, something funny happens. We are tempted when we try to talk about science as theory, to expect things to be different—that an independent reality will show itself. But closer attention to how people really work indicates a more complicated situation.

An example from my own experience: a significant part of my past graduate studies in engineering involved using X-rays to probe the structure of tiny solid crystals (Staley 1997). When the X-rays (a form of light, but an invisible one to us) pass through matter, their paths are diverted by interactions at the atomic scale, and a characteristic pattern results, so that the X-rays can "show" us how things are arranged inside the solid. This phenomenon (diffraction) is well known, and rigorous theories exist to predict that relationship: given a perfect crystal, as an ideal, we should be able perfectly to predict the pattern that ensues when perfectly aligned beams of light pass through them. Of course, nothing is perfect—there are other theories that tell us that that's impossible in our world. But the closer to perfect we get, the closer the prediction will match the evidence. And we know how to make instruments to create greater degrees of perfection, in controlling both the light and the solid structure. And greater perfection in the structure cashes out as functionality, which offers us a goal.

So the practical demonstrations that matter in this case tend to involve comparisons of what we expect to what we observe: Superimposing image and logic, as it were. Describing it in terms of our actual efforts, we make a picture of an ideal (graph the theoretical pattern), and we make a picture of the evidence (graph the data set), and we decide whether or not, or to what degree, they correspond. This, of course, is all heavily mediated by technology—the X-rays must be captured by a detector, which also filters things, and then must be made visible by turning the strength of that detected signal into a blip or trace, the theory is turned into a coded program that exists within a computer, where the pictures appear on a screen, and so on. Those in themselves are significant enough complications, and keeping track of all the mediations goes a long way toward making sense of things.

But we can go further and impose more "logic." Using statistics, we can describe quantitatively how closely the pattern and data match. Then, the computer can "decide" for us how good the match is. And the computer doesn't need to "see" the match. On the computer, there's no image. Nor does the computer have a goal. The pattern that we use to interpret things becomes irrelevant, and instead, what matters is a number (a statistic that describes the correspondence between two sets of other numbers). Or rather, a large set of numbers that the computer sifts through in order to "find out" how many can't be ruled out. Because in that world of logic, any result that's close enough to the perfect prediction, given the imperfections in the technology that's mediating everything, is a possible truth. It isn't that the two results are inconsistent—the human eye will happily agree with the pattern that corresponds to the computer's "best" answer, as well as many other "suboptimal" ones—but rather that they show an ineliminable ambiguity that resides within the image-logic divide. In Galison's story, the emphasis was on the (often idiosyncratic) language we use to negotiate those ambiguities when we encounter them. I take it that the complementary imperative stemming from Pitt's notion of technology as "humanity at work" involves focusing back on the various practical mediations from which such language flows. In this very grounded sense, the philosophy of technology then begins to look distinctly like a phenomenological project.

Parallel to this, I have argued elsewhere that talking about visual technologies in terms of images and seeing may sometimes profoundly disguise the details of human experience (Staley 2008). In that case, the specific issue at hand involved public understanding of new developments in nanotechnology. Alongside much hype at the end of the twentieth century about the alternately miraculous and devastating potential of controlling materials down to the atomic scale, there also came vivid imagery advertising our new capability of "seeing" at the atomic scale. Again, there's some sense to this claim, insofar as acting and observing are inextricably intertwined in the technological project here (to push the atoms around effectively, you must somehow know where they are and where they go). But the profound asymmetry between the phenomena relevant at the nanoscale and those we take for granted in our human meso-world also gives the lie to any claim of "seeing" atoms in a meaningful phenomenal sense. We just can't "be there" in the same way—not even in the way we can "be" outside the Milky Way looking down, where erasing distance by magnification is mostly what matters. There's no simulacrum we can inhabit that lives among the atoms. Instead, I think, when we talk about seeing atoms, we ought to discard the notion of "images" for that of "maps"—a different mode of representation where we recognize explicitly the partial nature of our access to the relevant reality and the category mistakes (intentional or not) involved in translating between our scale and that in which atoms operate. This problematizes the notion that "seeing" amounts to believing when we look with the technologized eye. Pitt might even argue that such depictions are mere "imaginings" (Pitt 2005). In either case, what's at stake is how to keep technoscience on a straight and narrow path toward the production of reliable truths—a problem in both epistemology and ethics.

Of course, the human experience is more than visual. These explorations of image and logic, the respective limitations of the two, and the prospects for technologizing our world can be productively broadened to include our other modes of sensation. For example, if we turn instead to hearing (Pitt and I both being enthusiasts of music), we can excavate a rich technological past in which the same fuzzy "image-logic" dipole transects the sounds that we hear and the musical, psychoacoustic, and physical systems that we use to describe them. These histories of the technologized ear may help us move away from the prejudice that it's through vision that our experience and our knowledge are primarily created.

Still further, we think with our noses too (Pitt and I both being enthusiasts of food, for example), and the technologized nose has its own story to tell. In the world of chemical sensation, there is a rich panoply of "images" written on our bodies by molecules powerfully enough that we can re-conjure our childhood through certain scents, or blindly recognize our lovers by apparent intuition. It's estimated that we each can distinguish thousands of distinct odors almost automatically, and of course, canines massively surpass our own capacities in this regard (Pitt being an enthusiast of dogs). The corresponding "logics" that we've created aim toward practical questions in many technosciences: The nature of molecularity; the medical efficacy of substances; the possible existence of "primary" smells or natural categories of odor stimuli. We make machines to measure scents, to concentrate them, and even to generate new ones that have never before existed. Industries have been built on perfume and empires based on spice. Here, I suggest, we might catch a good whiff (if not get a good glimpse) of a possible science of the passions. Many aspects of this history deserve to be better known, but here is not the place to elaborate on them. The philosophical point here, perhaps, is simply that decentering a default expectation (science is visual, vision is objective, seeing is believing) or extending a premise (the image-logic dipole) to its furthest extreme can often bear unexpected rewards—especially when attempting to grapple with human nature.

Human nature. Here we are, back around the table with David Hume. Pitt harbors a deep interest in Hume as an incisive analyst of human experience and a formative sociopolitical theorist. We moved slowly through *A Treatise of Human Nature* with him. The broad outlines of Hume's development there are fairly simple to articulate: the premise that our patterns of thought are all built on prior experiences ordered in space and time, according to varying degrees of resemblance—the epistemological arguments against both our access to causality and the logical notion of a self beyond our perceptions; the depiction of humans as ultimately ruled by passion rather than reason; the explanation of our social proclivities developing through forces of sympathy and association. (And, of course, there's more than that between the covers.) These matters all provoked serious debate in the room. But what stuck with me was primarily the issue of how to deal with Hume's perspective from a pragmatic standpoint.

What questions appear when we take Hume on pragmatically, on both philosophical and historical terms—considering the practical implications of his claims, their relation to and influence on the practice of others, and the ways in which Hume and those associated with him did their work as philosophers? Sometimes the seed of a dissertation can come from a single sentence, and in this case the seed was Hume's summary dismissal of serious attention to our senses: "The examination of our sensations belongs more to anatomists and natural philosophers than to moral; and therefore shall not at present be enter'd upon" (Hume 1738–40/1978, 8). This struck me as jarring. Hume hangs his whole conception of human nature on only three sorts of experience: contiguity, cause-and-effect, and resemblance. The first amounts only to proximity in space. The second he famously reduces to sequentiality in time. Only resemblance appears to offer any richness of experience, but that richness resides exactly in the broad spectrum of sensations that we are built to absorb. How can sensations be anything other than central to the task of examining human nature?

Now, the familiar reader might reasonably observe that Hume is after a different order of explanation. Since his ultimate aim (reached in Book III) is to consider the ways in which people operate as social creatures, the welter of sensations that we constantly collect is a level of minutiae that simply becomes a blur at the scale of our interpersonal relations. You don't need to see the individual frames to understand the film, as it were. Then too, for all the emphasis that's been placed on the Treatise as an epistemological text, Hume isn't trying to explain anything like the details of a process of technoscientific observation—his concern is for everyday experience. Further, Hume's self-identification as a moral philosopher is significant to his position, as it situates him in a particular tradition of his time, in which the division of labor within philosophy included an understood boundary between those of natural and moral stripes; after all, Hume isn't claiming that sensations are an unimportant concern, only that they are important in a different context.

But still: if we look at Hume's contemporaries and successors, we can find many moral philosophers who do worry about matters of sensation. Whether those figures identified Hume as an ally or enemy, or ignored him altogether, we find the attention of moral philosophers consistently returning to such concerns at greater and greater levels of detail as the process went on. And the enterprise morphed over time, while still dealing with the same fundamental matters as Hume: the individual and collective experience of the world in which we live our thoughts and passions. Where once stood moral philosophy, eventually emerged mental science, and ultimately a world of psychologists and sociologists and others. It was this history that I determined to explore, the better to situate Hume and his legacy in a practical context (for the outcome, see Staley 2004).

One aspect of that task was to compare Hume's express positions to various interlocutors—David Hartley, Thomas Reid, James Mill, Thomas Brown, William Hamilton, and Alexander Bain. Each of these figures developed positions that bear comparison to Hume, and with similar motivations. There are, in short, many resemblances. But writing a history of philosophical practice required moving beyond the book(s). As Hume himself could readily remind us, philosophy is a social enterprise, and so I came to emphasize—alongside the philosophical notions being promulgated in this tradition—a different set of associations: Who referred to whom and why? What was the membership of their social networks, and how were they organized? Where and how did they do their work? Such questions opened up a different kind of history of philosophy and disabused me of some significant preconceptions.

Instead of finding Hume (notorious provocateur and atheist that he was) as a key target for critics in succeeding generations, I encountered instead a surprising silence. It appeared that the flag of Associationism in the eighteenth century was carried instead by his more mystical contemporary, the physician David Hartley, who superimposed his own blend of Newtonianism and Christianity on the subject. The key role of editors came to the foreground when I found the second edition of James Mill's *Analysis of the Phenomena of the Human Mind* (1868) had been systematically doctored to replace the words "philosophy" and "philosophical" with "science" and "scientific"—these changes being instituted by his son, John Stewart Mill, and the latter's close associate, Alexander Bain. Bain, a figure now little remembered in the history of philosophy, himself produced two major volumes—*The Senses and the Intellect* (1855) and *The Emotions and the Will* (1859)—that, at great length, examine exactly the details of sensation that Hume chose to ignore. These works, explicitly intertwined with his close friend J. S. Mill's *System of Logic*, were a key touchstone for Victorian psychology in establishing Bain within a community of thought that coalesced significantly upon Bain's establishment of the journal, *Mind*, in 1876. Along the way, as I have already advertised, the enterprise of moral philosophy had bifurcated: much as the natural branch of philosophy began to shed new special sciences in the wake of Newton, moral philosophy began to breed special sciences of mind and behavior. The remaining rump of philosophy—as exhibited in the early pages of *Mind*—came to an apparent crisis. Philosophy, it began to appear, would have less leverage over the domain of human nature in a world populated by experimental psychologists and their ilk. Philosophers were seen less in the public sphere and more in academia. Informal collectives like the "Sunday Tramps" led by Leslie Stephen, who would sometimes descend without notice on Charles Darwin's home for lunch and conversation, were increasingly replaced by formal societies. Within philosophy itself, controversies over the proper orientation of work had begun to push apart an emergent school of logical empiricism from what became known as (mere?) speculative philosophy. And simultaneously, Hume resurfaced in the discussion (ironically resuscitated by his critic, T. H. Green). All of these developments represent aspects of philosophy grounded in daily practice and social stimuli. They are also indicative of a period of transition in which philosophy emerged significantly transformed by a cascade of (roughly Kuhnian) paradigm shifts (Kuhn 1962). Let's call that Trajectory 1, from Hume to Bain.

Especially striking to me in this study was a close reading of the inaugural issues of Bain's *Mind*. Trajectory 2 begins here. As the first formal journal of mental philosophy in the English language, this publication represents a key step in the professionalization of modern philosophy (Staley 2009a). It also serves as a tidy capsule of the issues at the forefront of the transition from moral philosophy to mental science (and the corresponding diminution of intellectual scope of philosophy at large). Here, in particular, I discovered another forgotten figure—Shadworth H. Hodgson—who proved surprisingly central in his time. Among other roles, Hodgson served as leader of a prominent discussion group known as the "Scratch Eight" (and who dominated the early tables of contents of Mind), maintaining a lengthy and influential friendship and lifelong correspondence with William James, and was the first president of the Aristotelian Society. On his retirement from that role, his successor eulogized his works as the "foundations on which the English philosophy of our hopes must one day be erected" (Bosanquet 1912, 2).

Hodgson's earliest contribution to *Mind* was a three-part exploration of the boundaries of "Philosophy and Science" (including "I: As Regards the Special Sciences," "II: As Regards Psychology," and "III: As Ontology"—Hodgson 1876a, 1876b, 1876c, and see Staley 2009b). This series articulates a position centering on what Hodgson calls the "two aspects theory" of human experience. Defending the notion that science and philosophy serve complementary functions in producing a "conspectus of reality," he works through a series of contrasts to identify distinctions among these enterprises. Philosophy, in particular, is conceived as offering an "ultimate subjective analysis of the notions which to science are themselves ultimate"—an organized depiction of the lived human world. Here, and even more fully in his later magnum opus *The Metaphysic of Experience* (1898), Hodgson develops an argument that the proper mission of "constructive philosophy"—the practical arm of the enterprise—is to provide a home for moral, aesthetic, and other human valuations. He finds this mission especially critical in the face of the competition for intellectual authority offered by the sciences-at-large, by psychology in particular, and by the philosophical specialty of epistemology (a term just then reaching currency).

These concerns fit neatly into a significant movement in English-language thought that was known collectively as "Speculative Philosophy," usually by contrast with "Scientific Philosophy" (as represented especially by logical empiricism). Between the middle of the nineteenth century and the middle of the twentieth, the defenders of speculative philosophy included (besides Hodgson) the influential American educator William T. Harris and his associates among the "St. Louis Hegelians," as well as the prominent British academics Alfred N. Whitehead, C. D. Broad, John T. D. Wisdom, and Henry H. Price (all, by the way, continuants in the Aristotelian Society and the work of *Mind*). The unifying motivation of these thinkers was to retain a place within philosophy—and science—for overarching human axiological concerns: "Imagination [as] a way of illuminating the facts" (Whitehead 1929, 93). Being "forced to look at the world synoptically" to avoid a narrow worldview "to which the natural scientist is peculiarly liable" (Broad 1924, 98). "Speculating and analyzing are operations which differ in kind; the object of one is truth, the object of the other is clarity" (Wisdom 1934, 1–2). "What the consumer mainly needs, I think, is a Weltenshauung, a unified outlook on the world" (Price 1963, 35). As practitioners, the speculative philosophers sought to educate people to negotiate the world, not to know facts.

And this brings us finally to Wilfrid Sellars. The second trajectory I've just been describing (Trajectory 2), with a succession of philosophers returning repeatedly to the question of how to connect our knowledge to our values, reaches its apogee with *Science, Perception and Reality* (in part, because Sellars ultimately knocks them down). Note the lack of the Oxford comma. I take it that Sellars isn't really talking about three distinct things. Rather, he is contrasting our perception of science with the reality of it. The collection begins with "Philosophy and the Scientific Image of Man," where he re-couches the project in terms that distinctly echo Hodgson and others in the speculative school. In this essay, Sellars develops a myth (an "idealization" in Sellars's terms) in which the tasks of philosophy and science appear almost as Arthurian quests (Pitt being an enthusiast of Arthurian quests). Roughly speaking, Sellars's "manifest image" encompasses our experience as we live it, while the "scientific image" extracts instead—through epistemic systems and technical mechanisms—classifications and explanations in logical terms. Being a realist about both aspects equally, Sellars is bound to recommend a "synoptic" or "stereoscopic" view. To quote his conclusion at length:

> Thus the conceptual framework of persons is not something that needs to be reconciled with the scientific image, but rather something to be joined to it. Thus, to complete the scientific image we need to enrich it not with more ways of saying what is the case, but with the language of community and individual intentions, so that by construing the actions we intend to do and the circumstances in which we intend to do them in scientific terms, we directly relate the world as conceived by scientific theory to our purposes, and make it our world and no longer an alien appendage to the world in which we do our living. (Sellars 1963, 40)

This reconciliation of intentions and facts offers us the prospect of a fundamentally "real" depiction of human experience while emphatically rejecting the prospect of ever finally getting to "the bottom of things." Our own comprehensively mediated access to the world (mediated by senses, by instruments, by language, by values, by history) is the ground on which both of Sellars's images develop, and it is through attention to such mediation that philosophy progresses. To the extent that things do ultimately hang together, we do justice to them in this way. Pitt's closely adjacent view (which he advertises as Sicilian realism—Pitt 2000) amplifies this by making explicit the multiplicity of identity intrinsic to all such experiential investigation—making hybridity the norm and purity the asymptotic exception so that we begin philosophy always with the expectation of complexity.

Note too that the situation as Sellars describes it above has a different tone than the one adopted by earlier generations of speculative philosophers, despite the similarity in the basic framework. Indeed, as I understand it, Sellars's account tidily pegs Hodgson and his fellow travelers as representatives of a "Perennial Philosophy" that ultimately privileges the manifest image—although examining that argument fully would divert attention from my main point here (but for more on perennial philosophy, see Pitt 2003). Suffice it to say that the position we are brought to at the end of "Philosophy and the Scientific Image of Man" is one that rejects both the primacy of either (manifest or scientific) image and a dualistic coupling of the two. At the end of the day, there is only one "real" for Sellars, but he reminds us too that we are not yet at the end of the day. Our discomfort at the apparently irreconcilable differences between the manifest and scientific frames is, on his analysis, a consequence of our historical moment, wherein the scientific image still requires completion.

Thus, rather than worry about the prospects for philosophy remaining relevant by distinguishing itself from science to provide "ultimate subjective analysis" (per Hodgson noted above), here we find an appeal from philosophy to strengthen science further. Like Hodgson, Sellars may conceive of the enterprise of philosophy as unique in its holism—the notion of "synoptic" or "stereoscopic" surveying that we've seen various practitioners advocate—but the survey that Sellars has conducted indicates a path for science to broaden its scope toward a more comprehensive form of realism. Continuing the visual metaphors, we might encapsulate this strategy as "seeing through people to study the human world."

Here's what I mean: as Sellars articulates it, the developing manifest image literally creates important aspects of reality. Specifically, the world of persons "comes to include" (in the mythically sequential sense in which one person's encounter with the world stands in for the perennially cyclical way in which we collectively encounter it) "us." That is, one consequence of the way that persons encounter the world is the existence-in-process of social phenomena. And in a fundamentally practical, rather than metaphysical, way these phenomena transcend and shape us. They exist independent of any given person, and by necessity, among all persons. Herein reside aspects of the manifest world that are accessible, but accessible only through persons in an instrumental manner—persons are the instruments by which we examine them, but they exist beyond. Sellars mentions among these interpersonal realities such things as communal intentions, duties, and meanings. Given their reality, and given our access to them, the appropriate philosophical strategy thus emerges not as competition with science, but in cooperation with better sciences: scientizing the manifest image and its phenomena, and taking those latter seriously as equally real with the "natural" ones we find around apparently independent of us. This requires the extension of the scientific image and the creation of new kinds of science that can adequately address the inherently social domain as it manifests itself through us: seeing through people to study the human world.

It may be useful here to mention another of Pitt's favorite reference points: C. P. Snow's classic exploration of the historical divergence of humanistic and scientific disciplines, *The Two Cultures* (1959). This was the first reading recommendation Pitt ever made to me. Much like the speculative philosophers who were his contemporaries of the last century, Snow is concerned to maintain a holism in the face of increasing specialization both in science and in humanistic inquiry, and to keep a conversation going across those boundaries—whether a literal conversation across the table at a Cambridge college dinner, or a figurative conversation at the global level. Snow seeks a "third way." But as more recent works like Jerome Kagan's *The Three Cultures* (2009) have suggested, it may be that—in Snow's moment—he missed a "third thing." This third thing, which Sellars has identified for us, is the potential for sciences of the human and the social to revisit our world as we live it and thus to bolster the connections that Snow found dissipating in his academic enclave. In a historical situation where philosophy's appeal to let the human back into the world of science has become commonplace where it was once novel, Sellars turns the question around to ask how best to let the world of science back into the human.

I'm well aware that this survey has run the risk of covering too much ground too quickly to capture everything that matters. The reader who wants a serious review of the many historical and conceptual details I've alluded to will need to look elsewhere—to the sources themselves, for a start. Then too, as Pitt has recently reminded us, complexity is what we should always expect. In that sense, I may have been speaking with a Sicilian accent here. But since my object has not been to illuminate ideas as much as practices, I hope we can now identify some worthwhile connections. We began with three disparate books, each of which was used by Pitt as a philosophical tool to stimulate conversation and thought. That was twenty years ago.

Since then, those stimuli have prompted me to explore questions about the ways in which we parse the world both individually through our senses and collectively through the language of perceptual qualities. I've thus done a lot of work trying to understand where the humanity lies in our instruments. I've also tried to find my bearings in the history of philosophy by thinking of the enterprise as a kind of productive work, and thus focusing on how and why ideas were developed and deployed, who used them, in what circumstances (rather more than the ideas themselves). And to the extent that my position within a university environment has provided me sufficient leverage, I have tried to facilitate the reweaving of intellectual fields that seem to be necessary in our current state of affairs—in my teaching, in the opportunities I have had to coordinate interdisciplinary research, and in my own attempts to remain acquainted with things beyond the piles on my desk.

Like many others Joe has taught over the years, what I do on a daily basis doesn't often look like philosophy (for my part, my current job description centers on teaching and management in a College of Engineering), but I hope that this essay will serve as some evidence that practicing philosophy isn't only for capital-P philosophers. But for someone like Pitt, who is a capital-P philosopher, perhaps we should go a bit further. I've described some of my own work process here, and I've indicated that I'm concerned more about matters of practice than (mere?) ideas. I've also suggested that if I and my fellows among Pitt's former students were to put our heads together on such considerations, the picture we could develop would have far more scope and clarity than the one I've sketched out alone. But what has Pitt been up to? Obviously, teaching us these things, and in retrospect I think teaching them to us rather well. And Pitt's extensive body of writings can speak for themselves as another obvious product that may be more than mere ideas.

But let's not forget the real work: Among Pitt's lasting achievements are the development of a department of philosophy from the ground up, from one-man shop to a diversified and vital enterprise that has influenced thousands—serving as department chair for more terms than he likely prefers to count. The development of two deeply interdisciplinary graduate programs (Science and Technology Studies, and ASPECT—the Alliance for Social, Political, Ethical, and Cultural Thought) that, not coincidentally, aim to inject strong contributions from social science into today's divide between humanities and the technosciences. Key roles in fostering organizations like the Society for the History of Philosophy of Science (whose first formal meeting he organized), and publications like *Perspectives on Science* and *Techné: Research in Philosophy and Technology* (for both of which he was long-time editor in chief). And the farm. And the dogs. And the Steinway. And, of course, Pitt's influence has been literally instrumental to the account I've developed here. And to many more stories. If you're a pragmatist philosopher, you must practice (how else could you get to Carnegie Hall?). And that's what philosophical work looks like.

Joe Pitt. Human at work.

NOTE

1. For the sake of formality, I will try to stick to calling our subject "Pitt" in what follows here, but the reader should recognize that to all of us who have worked with him closely he is always simply "Joe." Apologies for any slips in that regard.

REFERENCES

Bain, Alexander. 1855. *The Senses and the Intellect.* London: John W. Parker and Son.
Bain, Alexander. 1859. *The Emotions and the Will.* London: John W. Parker and Son.
Bosanquet, Bernard. 1894. "Presidential Address." *Proceedings of the Aristotelian Society*, Old Series. 3(1) 3–12.
Broad, C. D. 1924. "Critical and Speculative Philosophy" in *Contemporary British Philosophy: Personal Statements*, edited by J. H. Muirhead, pgs 77–100. London: Allen & Unwin.
Galison, Peter. 1997. *Image and Logic: A Material Culture of Microphysics.* Chicago: University of Chicago Press.
Hodgson, Shadworth H. 1876a. "Philosophy and Science I: As Regards the Special Sciences." *Mind.* 1(1) 67–81.
Hodgson, Shadworth H. 1876b. "Philosophy and Science II: As Regards Psychology." *Mind.* 1(2) 223–35.
Hodgson, Shadworth H. 1876c. "Philosophy and Science III: As Ontology." *Mind.* 1(3) 351–62.
Hodgson, Shadworth H. *The Metaphysic of Experience.* London: Longmans, 1898.
Hume, David. 1739–40/1978. *A Treatise of Human Nature*, edited by P. H. Nidditch. Oxford: Oxford University Press.
Kagan, Jerome. 2009. *The Three Cultures: Natural Sciences, Social Sciences, and the Humanities in the 21st Century.* Cambridge: Cambridge University Press.
Kuhn, Thomas S. 1962. *The Structure of Scientific Revolutions.* Chicago: University of Chicago Press.
Mill, James. 1968. *Analysis of the Phenomena of the Human Mind*, edited by J. S. Mill, Alexander Bain, and George Grote. London: Augustus M. Kelley Publishers.
Pitt, Joseph C. 1992. *Galileo, Human Knowledge, and the Book of Nature.* New York: Springer.
Pitt, Joseph C. 2000. *Thinking About Technology: Foundations of the Philosophy of Technology.* New York: Seven Bridges Press.
Pitt, Joseph C. 2003. "Against the Perennial: Small Steps Toward a Heraclitean Philosophy of Science." *Techné.* 7(2) 57–65.
Pitt, Joseph C. 2005. "When Is an Image Not an Image?" *Techné.* 8(3) 24–33.
Pitt, Joseph C. 2011a. "The Epistemology of the Very Small" in *Doing Philosophy of Technology: Essays in a Pragmatist Spirit*, pgs. 189–98. New York: Springer.
Pitt, Joseph C. 2011b. "Discovery, Telescopes, and Progress" in *Doing Philosophy of Technology: Essays in a Pragmatist Spirit*, pgs. 85–94. New York: Springer.
Pitt, Joseph C. 2019. *Heraclitus Redux: Technological Infrastructures and Scientific Change.* New York: Rowman & Littlefield International.
Price, Henry H. 1963. "Clarity Is Not Enough" in *Clarity is Not Enough: Essays in Criticism of Linguistic Philosophy*, edited by H. D. Price. London: Unwin Brothers.
Sellars, Wilfrid. 1963. *Science, Perception and Reality.* Atascadero, CA: Ridgeview Publishing.

Snow, C. P. 1959. *The Two Cultures*. Cambridge: Cambridge University Press.

Staley, Thomas W. 1997. "Statistical Optimization and Analysis of X-Ray Rocking Curves: Application to Epitaxial Alloys of Silicon, Germanium, and Gallium Arsenide." PhD diss, University of Wisconsin–Madison.

Staley, Thomas W. 2004. "Making Sense in Nineteenth Century Britain: Affinities of the Philosophy of Mind, c.1820–1860." PhD diss., Virginia Tech.

Staley, Thomas W. 2008. "The Coding of Technical Images of Nanospace: Analogy, Disanalogy, and the Asymmetry of Worlds." *Techne*. 12(1) 1–22.

Staley, Thomas W. 2009a. "The Journal *Mind* in Its Early Years, 1876–1920: An Introduction." *Journal of the History of Ideas*. 70(2) 259–63.

Staley, Thomas W. 2009b. "Keeping Philosophy in Mind: Shadworth H. Hodgson's Articulation of the Boundaries of Philosophy and Science." *Journal of the History of Ideas*. 70(2) 289–315.

Whitehead, Alfred N. 1929. *The Aims of Education, and Other Essays*. New York: Simon & Schuster.

Wisdom, John T. D. 1934. *Problems of Mind and Matter*. Cambridge: Cambridge University Press.

Chapter 8

Moral Values in Technical Artifacts

Peter Kroes

In the debate about values and technical artifacts one crucial question is not (sufficiently) addressed, namely, what is it that makes an object a technical artifact, for instance, a gun? Without a clear answer to this question the whole issue about values and technology cannot be addressed in a meaningful way. To illustrate this, I take as my lead Pitt's defense of the value neutrality thesis of technology. I agree with Pitt that technical artifacts are not the kind of things that act or do things. But I will argue that given the kind of things they are, they may bear values, albeit sometimes in a rather concealed way, and that, if we want to "see" them, we have to look for them in the right place.

I start with framing the issue under discussion in a very broad, rough way: in the following, I assume that there are no moral values in the physical and chemical world (nature), that is to say, the physical/chemical objects, considered by themselves, have no intrinsic moral values. The same is true for nonmoral values. Pieces of gold or diamonds may have a certain economic or sentimental value, but only so in relation to human beings. In physical theories there is no room for attributing any kind of value as an intrinsic property to physical objects. This is not a very controversial assumption. I also assume that there are no moral values, and for that matter no values at all, in the biological world. This may be a bit more controversial because of the use of notions like "survival value" of certain biological traits and discussions about intrinsic (moral) values of biological species, but it is broadly accepted that evolution theory has expelled values from our conception and description of biological nature (one of the key notions in evolution theory is the selection *of* trait X as opposed to selection *for* Y where one could substitute some value for Y). All in all, there are no values, in particular, no moral values in nature (see, for instance, Davies 2001), at least as long as we leave human beings out of consideration or conceive of human beings as a purely biological species on a par with other biological species. Values, including moral values, only come into play in relation to human beings when they are taken to be more than just physical, chemical, or biological entities.

What it is that somehow sets human beings apart from the value-empty world of nature I will leave open. Whatever it is, it will be closely related to the distinction between the natural and the artificial world, a distinction that plays a basic role in (Western) thinking about humans and their place in nature. The human world is indeed a world full of values. Human beings by themselves, in contrast to natural objects, have intrinsic moral value and may attribute all kinds of values to natural objects, but the values of these natural objects are, so to speak, in the "eye of their beholder."

Now, let's move from the natural and human world to the artificial world, more in particular: to the technological world which is a world of human making. What about moral values and technical artifacts? Are there moral values in technical artifacts considered by themselves? Clearly, technical artifacts, just as well as physical objects, may have values in relation to human beings. But does it make sense to assume that technical artifacts *may embody* moral values by themselves? That is the issue that we will concern ourselves with in the following pages.

To be clear, the issue is not about moral values and the *use* of technical artifacts. It is not controversial to claim that technical artifacts may be used in morally good or bad ways. Neither are we concerned with issues about the instrumental value of technical artifacts; it does make sense to claim that a technical artifact is good or bad instrumentally. The problem is whether it makes sense to claim that a technical artifact is morally good or bad by itself, independent of its instrumental goodness or of how it is actually used. So, do technical artifacts, just as human beings and their acts, belong to the category of things that may be morally good/bad per se?

In recent times the issue of the moral status of technical artifacts has received a lot of attention within the philosophy of technology, especially in relation to the so-called value neutrality thesis (VNT) of technical artifacts (see, for instance, the various contributions to Kroes and Verbeek 2014). This thesis states that technical artifacts by themselves have no moral values attached to them; they are instruments that, as such, have no moral significance. According to this view, it is simply a category mistake to attribute moral values (moral goodness or badness) to technical artifacts by themselves. This idea is exemplified in the slogan of the National Rifle Association (NRA): "Guns don't kill people, people kill people." Critics of VNT argue that technical artifacts are more than morally neutral instruments and that they, by themselves, come with some form of moral significance, which may be even some form of moral agency.

The NRA slogan illustrates that the actual relevance of our issue stretches far outside academic circles. Moreover, the history of this issue goes back far in time. It frequently plays a role in the background of age-old discussions about the moral impact of technology on human beings and about how to assess the role of technology in relation to what is called "the good life." In these discussions not only is the use of technology, but also technology itself, often attributed, often implicitly, some moral significance.

The gun has become probably one of the most widespread examples of a technical artifact used in discussions of VNT. For two reasons I find this example rather unfortunate. First, this example has become highly politicized due to the gun control issue in the United States. So a discussion of VNT may easily become tainted with political or ideological issues. Second, and much more importantly, it is not difficult to imagine situations in which guns may be designed, produced, and used (as guns) for morally good reasons (protection from violence), morally bad reasons (harming innocent people), or both. Therefore, it is not so obvious how guns, by themselves, may be taken to embody moral values, or, if they do so, then it seems that we are led to the rather uninteresting conclusion that guns by themselves may embody morally good and morally bad values at the same time. As I will argue in more detail later on, the gun is not a very suitable example for analyzing the moral status of technical artifacts.

I ask the reader to keep in mind in the following another (fictional) example, namely the execution machine as described by Kafka in his *In the Penal Colony* (Kafka 1919 [2015]).[1] This machine consists of various parts, including a bed on which a condemned person is tied and a part called "inscriber" with a harrow. It carries out the death sentence by inscribing the sentence with the harrow on the body of the condemned person. The machine is "not supposed to kill right away, but on average over a period of twelve hours." The following quote should suffice to give the reader a flavor of the horrific details of this machine:

> The harrow is starting to write. When it's finished with the first part of the script on the man's back, the layer of cotton wool rolls and turns the body slowly onto its side to give the harrow a new area. Meanwhile those parts lacerated by the inscription are lying on the cotton wool which, because it has been specially treated, immediately stops the bleeding and prepares the script for a further deepening. Here, as the body continues to rotate, prongs on the edge of the harrow then pull the cotton wool from the wounds, throw it into the pit, and the harrow goes to work again. In this way it keeps making the inscription deeper for twelve hours. For the first six hours the condemned man goes on living almost as before. He suffers nothing but pain. (Kafka 2015, 7)

What about the moral status of such a machine, if it would exist in reality?[2] Let me immediately add a caveat. Whatever emotional reaction this appalling machine may provoke, it will not play any role in my analysis. The reason why I prefer this example has nothing to do with emotions but with the intimate relationship between the function of this machine and particular human ends and values, a relation that is lacking in the gun example. I will argue that such a machine has moral significance by itself, independent of its actual use. However, how this claim is to be interpreted depends crucially, as we will see, on the meaning of the expression "a technical artifact by itself."

So, I will argue against VNT. As my starting point I take Joseph Pitt's (2014) defense of VNT in "'Guns Don't Kill, People Kill': Values in and/or Around Technologies." In my opinion it is the best, most detailed, and explicit defense of VNT available in the literature. His analysis will enable me to pinpoint precisely a crucial issue that, in my opinion, critics as well as defenders of VNT have generally ignored. It concerns the question of what makes an object a technical artifact. The answer to this question has far-reaching implications for how the expression "a technical artifact by itself" has to be interpreted in discussions of the moral status of a technical artifact. I proceed by first summarizing Pitt's main arguments for VNT. Then I present a critique of Pitt's defense in which I focus on Pitt's concept of a technical artifact. My next step is to propose an alternative according to which technical artifacts may have "inherent," but not "intrinsic" moral significance. In the final discussion, I briefly compare the moral status of technical artifacts with the moral status of human acts.

PITT'S DEFENSE OF THE VALUE NEUTRALITY THESIS

Pitt's basic position is that values "are the sorts of things that inanimate objects cannot possess, embody, or have" (Pitt 2014, 90). Technical artifacts are among those inanimate objects, and therefore, he is led to a defense of the VNT, which he formulates in the following way: "Technical artifacts do not have, have embedded in them, or contain values" (Pitt 2014, 90). I take this formulation of VNT to be equivalent to how I have stated this thesis above. One notable difference is my stress on technical artifacts *by themselves*. I assume that Pitt also has something similar in mind and that he would not deny that a technical artifact may have value *for* a human being, but it would be a mistake to conclude from this that this value is *in* the artifact itself.

Pitt is aware of the fact that his defense of VNT needs an account of values, a topic that I have not touched upon so far. However, he remarks that defining values may raise a serious problem since "I do not believe it is possible to develop such an account without begging the question, i.e., that values are the sorts of things that only humans have" (Pitt 2014, 90). This, he claims, raises a potential dilemma. If this is true, then, on the one hand, VNT is true by definition. So, if VNT is taken to be a rather strong claim with real content, then we would have no idea about how to justify it. On the other hand, it may be argued that since humans act and make decisions on the basis of values, anything made by humans, including technical artifacts, will be tainted somehow by human values. That will render VNT trivial. In his article Pitt focuses mainly on the second horn of this dilemma and defends, what he calls a weaker version of VNT, that "claims that even if we could make sense of the idea that technical artifacts embody human values, there are so many that would be involved the claim says nothing significant" (Pitt 2014, 101).

The account of values on which Pitt bases his defense of this weak version of VNT is a pragmatist one: "a value is an endorsement of a preferred state of affairs by an individual or group of individuals that motivates our actions" (Pitt 2014, 91). This, indeed, implies that only human beings have values. He stresses that values are not goals, which can be achieved, but action initiators that guide our actions in certain directions. Whatever the merits of this pragmatist account of values may be, I will follow Pitt and take it also as my starting point. For my critique of VNT it will not be necessary to present a detailed account of values. What is important for the following is that I agree with Pitt that values, however they are construed, are the sorts of things that only human beings have. Nevertheless, I will defend the claim that technical artifacts may "have" values in a meaningful, not just metaphorical, way. As Pitt remarks, to interpret VNT such that technical artifacts have a kind of value sui generis runs indeed the risk of begging the question we are dealing with (Pitt 2014, 96).

With his account of values in place, Pitt rests his defense of VNT on two main arguments. First, he claims that technical artifacts lack empirical features that make it possible to identify their embedded values. Second, the making of technical artifacts involves many decisions, and these decisions in turn involve numerous values. As a result, if technical artifacts are somehow tainted by these values, the "final conclusion is going to be the claim that so many values are involved in the creation of an artifact that we might as well say it is value neutral" (Pitt 2014, 91).

In his first argument, Pitt uses Langdon Winner's (in)famous example of Robert Moses's overpasses of the Long Island Expressway (LIE); according to Winner, Moses's (racist) values are embedded in these overpasses.[3] So Pitt asks himself where these values are to be found. They are not to be found in the design of the LIE overpasses: "Where would we see them? Let us say we have a schematic of an overpass in front of us. Please point to the place where we see the value" (Pitt 2014, 95). They are not in the lines, numbers, or any other sign in the blueprints. Likewise, they are not to be found in the actual overpasses: "if we look at the actual physical thing—the roads and bridges, etc. where are the values? I see bricks and stones and pavement, etc. But where are the values—do they have colors? How much do they weigh? How tall are they or how skinny? What are they?" (Pitt 2014, 95). Leaving Pitt's rhetoric aside, it is indeed difficult to point to empirical features of the design or the actual physical object on the basis of which it would be possible to identify the values, as initiators of actions to achieve preferred states of affairs, that are embedded in Moses's overpasses. Apart from locating where values are embedded in technical artifacts, whose values are embedded? Since many people may be involved in the designing and making of a technical artifact, this question may be equally difficult to solve.

The issue with whose values brings us to Pitt's second argument. Suppose that somehow values may be embedded in technical artifacts. Then it follows that, given the fact that usually many different people with different values are involved in making technical artifacts, a myriad of values will be embedded. The whole process of product development, from its first conception through design to making, involves a myriad of value-laden decisions and since all these decisions shape the final technical artifact, they all end up in the technical artifact. For Pitt, this means that too many values will be embedded to single out any one in a nonarbitrary way and that therefore the claim that technical artifacts have values becomes a trivial, uninteresting one. He sums up his defense of this weaker version of VNT as follows: "since artifacts are the results of human decisions and since human decisions are a function of human values, understood as motivators to achieve a certain preferred state of affairs, and since many people are involved in the creation of technical artifacts, it adds nothing to the discussion to say values are embedded in artifacts" (Pitt 2014, 101).

CRITIQUE OF PITT'S DEFENSE

I find both arguments problematic for reasons that concern an issue that Pitt does not address but that in my opinion ought to play a crucial role, next to the issue about values, in any discussion of VNT, namely, the issue of what kind of objects are technical artifacts. Although Pitt is clearly aware of the fact that a defense of the VNT needs an account of values, and discusses this issue in depth, he seems not aware that he also needs *an account of technical artifacts*. He does not explicitly address the question of what kind of objects technical artifacts are. Nevertheless, his first argument contains sufficient clues to reconstruct how he conceives of technical artifacts. Before taking a look at these clues, I will first comment briefly on his second argument.

Contra Pitt, I think it is possible to single out in a nonarbitrary way certain values that play a special role in the making of a technical artifact and that it may embody, in a way still to be explained. Of course, he is right in claiming that nowadays usually a multitude of people are involved in engineering practices of conceiving, designing, and making a technical artifact. All of them come with their own interests, values, and norms and make decisions on the basis of these that may affect the final outcome. Engineering practice is a social practice and as such is, indeed, thoroughly value-laden by all kinds of values, norms, and interests that guide the actions of individuals and groups of people.

So whose values and what values end up being embodied in the technical artifact? Well, not all of these values turn out to be on a par as soon as we realize that not any social practice is an engineering practice. According to Alasdair MacIntyre, a practice is defined by the specific internal goods it produces and its standards of excellence:

> By a "practice" I am going to mean any coherent and complex form of socially established cooperative human activity through which goods internal to that form of activity are realized in the course of trying to achieve those standards of excellence which are appropriate to, and partially definitive of, that form of activity, with the result that human powers to achieve excellence, and human conceptions of the ends and goods involved, are systematically extended. (MacIntyre 1984, 187)

If we take this conception of a practice as our starting point, then engineering practice is defined by the specific goods it produces and by its standards of excellence. One kind of internal goods produced by engineering practice is technical artifacts. A core feature of technical artifacts in general is that they have instrumental value, which brings into play two other values (with their corresponding standards of excellence): namely *efficacy* and *efficiency*. Technical artifacts in general are supposed to have instrumental value by being efficacious and efficient. More particular values may be related to technical artifacts by not focusing on engineering practice and technical artifacts in general, but by zooming in on specific engineering practices that develop specific kinds of technical artifacts, such as life vests, guns, knives, or Kafka's torture machine.

These different kinds of technical artifacts have instrumental value for realizing particular human ends that are in turn related to particular human values. Among all the values involved in engineering practice as a social practice, these values have a special status. They can be picked out in a nonarbitrary way because these values are intimately related to the kind of thing that is produced in a certain engineering practice, namely, a specific kind of technical artifact. These values lie at the basis of the core functional requirements that the technical artifact involved has to satisfy. And, of course, any specific kind of technical artifact has to satisfy also the instrumental standards of excellence of technical artifacts in general, that is, efficacy and efficiency. So, we are able to answer Pitt's what-values-question. The whose-values-question turns out to be not of much interest, since what kind of technical artifact is produced is determined, among other things, by what kind of human ends and values are involved (for which ends it has instrumental value), not by whose ends and values.

The foregoing still leaves us with the question of how a technical artifact may have these values. According to Pitt's first argument, it cannot have values because a technical artifact lacks empirical features necessary to identify the values embedded in it. Pitt looks for them in the empirical features of its design, which he takes to be its working drawings or blueprints or in the actual physical object of a technical artifact, but it is all in vain: nowhere any empirical clue is to be found as to where and which values are in the artifact. I have two worries about the way in Pitt deals with the issue of locating the embedded values.

The first one, on which I will only briefly touch here, concerns the fact that Pitt assumes that there is a rather direct link between values and empirical features. He defends a pragmatist account of values because it relates values to empirically observable actions (p. 94): "if you claim you have a certain value, then you must do something to show that you, in fact, endorse that value." But it may be questioned whether values in general are features of human behavior that are empirically identifiable in such a direct way. A lot of interpretation may come in, even on his pragmatist approach to values. From a behaviorist point of view, the notion of value may even be taken to be a highly theoretical one. So, to ask, as Pitt does, to locate them in directly observable features of technical artifacts may be asking too much.

Be that as it may, my second and most important worry concerns Pitt's concept of a technical artifact on which his defense of VNT is based. Apparently, for Pitt a technical artifact is nothing but a physical object. Looking for the values embedded in the LIE he writes: "Likewise for the LIE—if we look at the *actual physical thing*—the roads and bridges, etc. where are the values?" (Pitt 2014, 95, italics added). Even if we take a technical artifact to be a physical object together with its design, it ends up being some physical object since for Pitt a design appears to be just a set of physical drawings (blueprints). Given such a concept of technical artifacts, it comes as no surprise that we can't find any values in them. As I remarked at the very beginning, it is generally accepted that physical objects do not have values. So we have to face the question of whether Pitt's concept of a technical artifact as nothing but a physical object makes sense.

In my opinion, the question of what kind of objects technical artifacts are is as crucial for a discussion of VNT as the question of what values are. Unfortunately, this question is often not addressed in defenses as well as critiques of VNT. For various reasons, I think that the concept of a technical artifact as merely a physical object is inadequate and, therefore, also Pitt's defense of VNT. One of the main reasons is that it is difficult to make sense of evaluative statements about technical artifacts, such as the statement "This is a good/bad hammer"; a hammer, as nothing more than a physical object, cannot be a good or bad hammer, just as an electron cannot be a good or bad electron. From this perspective, the most obvious interpretation of the function of a technical artifact is that it is a physical capacity, but this interpretation is problematic: whereas it makes sense to claim that a function may be performed well/badly, this is not the case for a physical capacity. Apart from this, the interpretation of technical artifacts as physical objects has problems in dealing with the fact that technical artifacts come in different kinds.

Hammers are a different kind of artifacts from doorstops and a hammer, even when it is used successfully as a doorstop, is not a doorstop. How to explain all this? On the one hand, if physical properties (capacities) determine artifact kind, then why would the hammer not also be a doorstop, since it has the appropriate physical capacity for being a doorstop? On the other hand, if the use context determines the artifact kind, then there is also no reason to deny that the hammer is a doorstop. This issue about artifact kinds is closely related to the distinction between accidental and proper functions, a distinction that the conception of technical artifacts as mere physical objects has difficulties dealing with (for more details, see Kroes 2012, Ch. 2 and 4).

In the next section I will outline a view on technical artifacts that, in my opinion, does not suffer from these problems and according to which, in a sense yet to be explained, it may be claimed that, contrary to VNT, technical artifacts by themselves may have values.

TECHNICAL ARTIFACTS WITH VALUES AND INHERENT MORAL SIGNIFICANCE

In my view, technical artifacts are more than just physical objects; they are physical objects, usually human-made physical constructions, *with functions*. These functions confer a certain "for-ness" on technical artifacts: they are, *qua* technical artifacts, objects (instruments) for realizing certain ends. It is the function of a technical artifact that distinguishes it from being just a physical object. If we look at engineering practices of designing and making technical artifacts, but also their use practices, then it becomes clear that technical artifacts are more than just physical objects. With regard to a technical artifact, it always makes sense to ask at least the following three questions: (1) What is this object for? (2) What is this object made of? And (3) How is this object to be used? (Kroes 2012, 43). With regard to a physical object, the first and third questions do not make sense at all. Apparently, we are dealing here with two different kinds of objects: what makes technical artifacts distinct from physical objects are their functional properties. These functional properties are conceptually definitive and ontologically constitutive for being a technical artifact.

Of course, this view requires a further explication of what functions or functional properties are. As we have seen above, the functional properties of a technical artifact cannot be equated with (one of) its intrinsic physical capacities. Such a view on functions reduces a technical artifact again to just a physical object, and is unable to deal with normative claims about technical artifacts. Another way is to interpret functional properties only in terms of human intentions. According to this view, physical objects have functions (and are therefore technical artifacts) because humans have certain intentions (beliefs, ends, desires, etc.) about them. What is attractive about this view is that it makes it possible to relate the functions (for-ness) of technical artifacts to human ends, what is problematic is that the physical properties of a technical artifact appear to be irrelevant for its functional properties. It is obvious, however, that the physical properties of a technical artifact play a crucial role in performing its function; its functional properties are therefore intimately related to its physical properties. But the same is true for functions and human intentions: in the making and using of technical artifacts, human intentions, in particular human ends, play an essential role.

So what is needed is an interpretation of technical artifacts that ties their functional properties to their physical properties, on the one hand, and to human intentions, on the other. Whereas the physical properties are intrinsic, the functional ones are relational properties. The functional properties of a technical artifact cannot, therefore, be equated with some of its physical capacities because the latter are intrinsic properties. Technical artifacts have functional properties only in relation to human intentions. Human intentions relate technical artifacts as functional instruments to human ends and values. According to this view, technical artifacts have a dual nature: they have physical and intention-related properties. They are a kind of hybrid object between the world of physical objects and the world of intentional objects.

It is this hybrid nature of technical artifacts that, in my opinion, puts the VNT in a different light. Pitt's defense of the VNT is based on a conception of technical artifacts that takes them to be nothing more than physical objects. From that perspective, it is indeed difficult to make sense of the idea that technical artifacts may have or embody values; whatever values they would have or embody, they would be intrinsic to them and completely independent of human values and intentions. So, Pitt is driven to the conclusion that "values are the sorts of things artifacts cannot have in any *meaningful* way" (Pitt 2014, 91). However, by taking due account of their hybrid nature, we may conclude that technical artifacts by themselves can have or embody values in a meaningful way. But we should take care in interpreting the expression "technical artifacts by themselves." Given that technical artifacts are inherently related to human intentions, the expression "technical artifacts by themselves" may be rather misleading since it may be taken to mean that we can consider technical artifacts as objects independent of human intentions. However, technical artifacts *by themselves* are not objects with only intrinsic properties; that would lead us back to Pitt's conclusion. Technical artifacts are *relational objects* with relational (functional) properties, and these relate technical artifacts to human values. Since these relational properties are conceptually definitive and ontologically constitutive for being a technical artifact, technical artifacts are *inherently* related to human ends and values. The term "inherently" means that technical artifacts by themselves come with values attached to them, but these values are not intrinsic to them in the sense in which their physical properties are intrinsic. Some of these values may be moral values, and therefore, technical artifacts by themselves may have inherent moral significance.

So, if we want to look for values in a technical artifact, we should not examine its physical properties, but its functional ones; together with its physical ones, they define or constitute an object as a technical artifact. Actually, this conception of a technical artifact makes it easier to find a coupling between empirical features and values, a coupling that Pitt is so desperately searching for. We perceive many human-made material objects in our environment as physical objects with a function, that is, as technical artifacts and not as pure physical objects, that is, as objects with only properties that are countenanced by the physical sciences; a bathtub, for example, is not perceived as an object with only geometrical and material/physical properties, that is, as an object that is constituted by an aggregate of physical atoms, but as an object with functional properties. So, in perception, the functional properties of objects appear to be directly present to the perceiver.

To underscore this point, let me repeat the above quote from Pitt: "if we look at the actual physical thing—the roads and bridges, etc. where are the values? I see bricks and stones and pavement, etc. But where are the values—do they have colors? How much do they weigh? How tall are they or how skinny? What are they?" When looking at the actual physical thing, Pitt sees bricks, stones, and pavement; so he sees technical artifacts; he sees things that on top of their physical properties also have functional properties! But nevertheless, he does not see their values. Why? I think the reason is that he *thinks* that he is looking at physical objects and for him that are not the kinds of objects that may have values. But when looking at roads and bridges, we do perceive *technical* values that are intimately associated with functional properties of these objects that, in turn, are intimately associated with some of their specific empirical features. For example, the flatness of the pavement of a road is related to the value of efficiency; we do "see" that a certain pavement is bad for driving. We may even perceive moral values in case we "see" that a certain pavement is dangerous for driving. A much more clear-cut case of a technical artifact that bears its moral value much more in the open is Kafka's torture machine, because its overall function, as realized by its physical structure is inflicting as much pain as possible to a human being. In my opinion it is, in contrast to the gun, a clear example of a technical artifact with inherent moral significance.

DISCUSSION

It would be a mistake to conclude from the above analysis that *any* technical artifact has inherent moral significance. For Kafka's execution machine, that is the case because the ends implied by the function of this machine have a clear moral significance. But for many technical artifacts (such as screws, nuts and bolts, hammers, etc.) there is simply not such an intimate connection between function, human ends, and moral values. In this respect an analogy with human acts may be drawn. Some human acts, lying, for instance, are generally considered to be morally problematic by themselves, because they violate an important moral norm related to the value of truth. But many, if not most, acts lack such a clear connection to moral values and therefore have no moral significance by themselves. With regard to technical artifacts, guns may indeed lack inherent moral significance because there is no clear-cut relation with particular human ends and values. If so, then in this respect, guns are not representative of all technical artifacts and may therefore be ill-suited as an example for discussing the VNT.

It should also be noted that technical artifacts with inherent moral significance may be used in ways which are not in line with the moral values attached to them. Surely, just as Kafka could imagine a story in which the "injustice of the process and the inhumanity of the execution were beyond doubt" (Kafka 2015, 8) some other writer may be able to come up with a story in which this gruesome machine is used, even as a torture machine, in a way that may be morally acceptable. Here the analogy with acts also holds as, for instance, is illustrated by the notion of a "white lie."

Finally, my critique of VNT should not be misunderstood. According to Pitt, one reason why people would want to deny VNT has to do with responsibility (p. 96): "Why would anyone deny that technologies are value neutral?

One answer that presents itself is 'To escape responsibility for their actions.' To say that means something like this: the machine made me do it. And that is totally absurd. Machines don't make you do anything." I will leave aside whether it is totally absurd to claim that machines can make people do things, but I agree with Pitt that ultimately "[t]he culprit is people" (Pitt 2014, 101). I am not for attributing moral agency to technical artifacts, since that may turn us away from the responsibility we bear for the world of our own making. Whatever moral values are embodied in technical artifacts, they are embodied in them as the result of human action. In my opinion, denying VNT is compatible with an "ethical heuristics" that strives for an approach to the moral status of technical artifacts in which humans bear, as much as is reasonably possible, the moral burden of what we do to each other, with or without the help of technical artifacts (for more details, see Kroes 2012, Ch. 6).

NOTES

1. I thank Maarten Franssen for drawing my attention to this example.
2. Note that I will not be discussing here the issue whether a fictional entity may have moral significance.
3. I leave the issue of whether Winner's story about Moses is historically accurate or not aside. The issue under consideration is not a historic but a conceptual one: suppose that Winner's story is historically accurate, then what does Winner's claim that Moses's bridges have (racist) values embedded in them mean.

REFERENCES

Davies, P. S. 2001. *Norms of Nature*. Cambridge, MA: MIT Press.

Kafka, F. 2015. *In the Penal Colony*. Translator I. Johnston. Global Grey. www.globalgreybook.com.

Kroes, P. 2012. *Technical Artefacts: Creations of Mind and Matter*. Dordrecht: Springer.

Kroes, P. and P.-P. Verbeek (eds.). 2014. *The Moral Status of Technical Artefacts*. Springer.

MacIntyre, A. C. 1984. *After Virtue: A Study in Moral Theory*. Notre Dame (Ind.): University of Notre Dame Press.

Pitt, J. C. 2014. "'Guns Don't Kill, People Kill': Values in and/or Around Technologies." In *The Moral Status of Technical Artifacts*, edited by P. Kroes and P.-P. Verbeek. Springer, 89–101.

Chapter 9

Engineering Students as Technological Artifacts

Reflections on Pragmatism and Philosophy in Engineering Education

Brandiff R. Caron

Like Joe Pitt, it took me some time to come around to pragmatism.[1] For most of my studies and well into my early career as a philosopher of technology, I was possessed by an unshakeable sense that pragmatists were changing the rules of the game and claiming victory. Seemingly profound questions as to the nature of Truth, Reality, and Morality, questions which had motivated my interest in studying philosophy in the first place, have been formulated un-usefully and, therefore, inappropriately, my pragmatist friends argued. There is no difference between practical and theoretical reason, they would insist. Having used these aged, obsolete questions to climb up the ladder of philosophy, I must now see the senselessness of them and throw the ladder away. Whereof one cannot speak, thereof one must be silent![2]

My adherence to Bertrand Russell's conception of what "philosophy" is (or ought to be) did not sit well with the philosophical pragmatism I encountered. Russell maintained that it is not in spite of the uncertainty surrounding philosophical questions that their study retains value but rather because of it. As soon as something approaching certainty is attained in answering certain questions, that field of inquiry has matured into a science and moves out of the realm of philosophy. The philosopher then turns her attention away from this field of inquiry (except where conceptual muddiness requires philosophical clarification—that is, philosophy of science) and turns toward new spaces of messiness and uncertainty. In this view, a view I held for a long time, philosophers are, by definition, delegated the task of living in and working (only) with uncertainty. This has always appealed to me. It sits well next to my intuition that the world itself and our experience in it are indeed messy and uncertain!

This understanding of the role of the philosopher, naturally, creates a problem for those inclined to insist on the practical utility of philosophy. And it created a problem for me.

I didn't even have my PhD in hand (I had yet to defend my dissertation) when I accepted a temporary position as a lecturer in something called the "Center for Engineering in Society." The center was situated within the faculty of engineering and computer science in a large Canadian university. My first experience as a practicing, professional philosopher was not the typical experience most academic philosophers have. I was not to be surrounded by other philosophers who shared a basic understanding of the intellectual history and general trajectory of my scholarly and pedagogical interests. No, for me, it was to be surrounded by academics and scholars who, when I was lucky, had little understanding of what I did or how I did it. When I was unlucky, there was outright contempt for the liberal arts and social sciences that I came to represent. My teaching and scholarship had to become framed as answers to the question, "What can you do for us?"

My interests in democracy theory were funneled through pragmatic concerns surrounding how to include end users in open-ended engineering design projects. My interests in philosophical theories of ethics were funneled through pragmatic concerns surrounding professional codes of ethics. And my interests in pursuing a disruptive, critical pedagogy were funneled through pragmatic concerns surrounding accreditation criteria. The messiness and uncertainty that comprise the best theories we have to describe the relationship between science, technology, and society, insights that had heretofore been the very currency of my scholarship, needed now to be operationalized into practice somehow. I needed to show what I (qua representative of all liberal arts and social scientific scholars and educators) could do for them (all engineers and computer scientists) now.[3]

I inherited a syllabus for the first course I was to teach in my new position.[4] The syllabus was for a course called "Impact of Technology on Society." Grimacing as I read the title for the first time (on account of the fact that I knew that the relationship between technology and society to be far too messy to fit neatly into "impact studies"—had these people never heard of social constructivism!?), I was pleasantly surprised to see that the syllabus was very nearly identical to the one I had been using at Virginia Tech for the course, "Introduction to Science and Technology Studies." It was with great enthusiasm, then, that I encountered the few readings on this new syllabus that I had not already used in my teaching at Virginia Tech.

The course syllabus started off in the same way any good course with a (at least a partial) philosophical bent ought to: definitions. Definitions are important. This is especially true in a course designed to provide engineering students (what we like to refer to as "engineers-in-the-making") the tools to be able to reflect on the ethical, legal, social, and environmental aspects of the technologies they are learning how to create. Borrowing from the science and technology studies literature on Technology Assessment, we then provide students with the tools to be able to reflexively engage with and *alter* these aspects of engineering projects (when, say, an ethical or legal standard is found to be problematic in a design) in addition to simply reflecting on them.

I like the phrase "engineers-in-the-making." It reminds us that we are working with the people who are most directly involved in the emergence of new technologies (just a bit earlier in the timeline before they are in this position and, hopefully, while they are still impressionable enough to learn something about the nature of technology beyond what they are bombarded with in a traditional North American undergraduate engineering curriculum). I emphasize this point, because this idea, the idea that engineers can and ought to reflexively engage with the social and ethical aspects of their technical designs and prototyping—that every iteration of the R&D process can and ought to have this ethical reflexivity—is at the heart of this course. The working definition of technology in the course, therefore, must allow for and make full account of this possibility.

Engineering students often come into the class with a definition of technology already in mind. This definition is rarely a fully formulated one. Rather, it is something cobbled together and accumulated slowly through a regular implication of how "technology" gets framed and mobilized in their technical courses. Technology is artifact. Engineers work with objects, things. Artifacts, objects, and things that can and ought to be treated as independent of the ethical, legal, social, and environmental issues that reside in some other, far messier realm.

Having discussions around defining technology at the beginning of this course is necessary to combat this definition of technology, that, if true, would mean that engineers either do not have to consider the messy stuff in this other realm or, in addition to the cramped technical curriculum, have to become social scientists in addition to engineers. The former is unacceptable, given the aims of this course, and the latter is unacceptable to engineering curriculum coordinators (at least most of the ones I have met). The way to handle this is to show the ways in which technology necessarily, that is, by definition, has social aspects.

I had, before inheriting this new syllabus, often used Joe's chapter on defining technology from his *Thinking about Technology* (2000) in introductory courses to science and technology studies. In this chapter, Joe points out the need for philosophers of engineering (and, I take it, engineers themselves) to "rethink the old assumption that technology is merely applied science, and as such, applied knowledge" (Pitt 2000, 9). This old assumption, the tool-as-mechanical-mechanism notion of technology, is precisely what underlies the understanding of technology that undergraduate engineering students bring to this class. It is precisely what Joe wants to combat. It is precisely what I want to combat.

Technology must, instead, be understood to be a process with social dimensions (alongside, or in addition to, the technical or scientific). "If something can be used to achieve a goal, it is a tool and, in so being, can become a technology" (Pitt 2000, 9). This proposition contains the two ingredients that lead Joe to assign his infamous (I will explain the "in" in "infamous" later) definition to technology: technology is humanity at work (Pitt 2000, 9). It's worth foregrounding the two ingredients here: (1) it is the activity of humans that is the object of concern in defining technology and (2) it is humans' deliberate and purposeful use of tools that characterizes technology.

In the new syllabus, I was asked to use the concept of narrative to define technology as introduced in David Nye's *Technology Matters: Questions to Live With* (2006). The first chapter in this book, appropriately titled "Can We Define 'Technology?'" lays out Nye's understanding of technology as narrative. He does so in a way that is similar to Joe's proposition above. Nye claims that technology can be thought of as a narrative with a beginning, a middle, and an end. To explain what a tool is and how to use it demands a story. To get the reader's intuitions about the social dimensions of technology flowing Nye uses a thought experiment that I have found effective with undergraduate engineering students. He asks us to imagine we have locked ourselves out of our car and are considering using a rock on the ground or a twisted coat hanger to gain access to the locked car:

> There is a situation; something needs doing. Someone obtains or invents a tool in order to do it—a twisted coat hanger, for example. And afterwards, when the car door is opened, there is a new situation. Admittedly, this is not much of a narrative, taken in the abstract, but to conceive of a tool is to think in time and to imagine change. The existence of a tool also immediately implies that a cultural group has reached a point where it can remember past actions and reproduce them in memory. Tools require the ability to recollect what one has done and to see actions as a sequence in time. (Nye 2006, 11)

In this way, Nye gives us a three-part definition of technology that follows the structure of narrative. There is a beginning, a middle, and an end in what constitutes technology. We require a set of intentions (beginning), an artifact or tool (middle), and a set of desired results (end). Students often find the example of the rock on the ground the most convincing evidence that this definition is getting at something about the nature of "technology." We all agree that, prior to being picked up and used as a means of breaking a car window in order to acquire the keys locked inside, the rock on the ground is a mere artifact. It is neither a tool nor a technology. However, the moment we pick it up and plug it into this narrative structure, we would all agree that it is indeed a technology. This example shows that a *necessary* component of "technology" is a set of intentions (human or otherwise) and desires (human or otherwise) for a different future.

Nye's use of intentionality and desire track Joe's use of goal orientation in defining technology: "If something can be used to achieve a goal, it is a tool and, in so being, can become a technology" (Nye 2006, 9). In this language, we can say a *necessary* component of "technology" is a goal. I stress the notion of necessity here for the same reasons I do in class. It is important to see what we have done here.

Technology should be defined as something that is composed of (most often human[5]) intention.

If we wish to speak of neutral artifacts independent of the human sphere of intentions and desires, we ought not to use the concept of technology. I prefer the term "artifact" to "tool" for precisely this reason. "Tool" still carries with it a notion of the intentionality and goal orientation that is reserved for Nye's and Pitt's definitions of technology.

Technology = (Intention + Artifact + Desired Outcome).

An artifact alone, just as an intention alone, does not constitute a technology. This is not a simple semantic distinction. Engineers do not research and develop random artifacts with no connection to a goal. We cannot say, as so many engineering students in my courses do, that engineers are only responsible for the nonnormative artifactual component of technology. Given the discussion here regarding goals and intentions in the composition of "technology," this no longer even makes sense.

The next move I make in class, a move that is made by most scholars of science and technology these days, is to point out that since our concept of technology has, by necessity, a set of intentions and desires (in the form of goals) and since intentions and desires are by definition normative concepts (one can have good or bad intentions), technology itself can be (and must be) understood to be normative. By this reasoning, it is a mistake to claim that individual technologies are neither good nor bad. And yet, this is precisely the sentiment one finds in much of Joe's writing. Indeed, Joe seems to have a great distaste of thinking about technology through the lens of ethics, going as far as to accuse those that flirt with the idea, such as the co-authors of *Pragmatist Ethics for a Technological Culture* (Keulartz et al. 2002), of engaging in "ethical colonialism" (Pitt 2004, 32–38). Joe's primary objections to this line of reasoning are wrapped up in how one understands the "agency" of technology. The worry is that many who have tried to understand technology through a normative lens have attributed causal powers to it and endowed it with a mind and intentions of its own (Pitt 2000, 87).

What position on the agency of technology are we committed to when claiming technology is normative? Are we committed to ascribing a form of autonomy to technology if we claim it is normative? These are the questions that motivate much of Joe's work. In arguing for the position that technology is not an autonomous force independent of human will, a position shared by most STS scholars today, Joe takes on those that would argue that some technologies exert an inexorable will on society, forcing society to accept the normative ends of this technology. Are all positions that claim technology is normative committed to such a view on the agency of technology?

In what follows, I track how agency is attributed (or not) to technology across a spectrum of theories that try to explain the relationship between technology and society. In doing so, I hope to show the availability of theories that allow for appropriate ethical evaluation of technology without ascribing autonomy to technology.

After answering, "Can we define technology?" Nye moves next to another question: "Is technology inherently deterministic, or is it inflected or even shaped by culture?" In his second chapter, titled "Does Technology Control Us?" Nye describes theories of technological determinism. He does so in a way that tries to impress upon students a historical understanding of these theories and the contexts in which they have been taken seriously, even forming common-sense understandings of the relationship between technology and society. However, Nye quickly dismisses these theories and devotes the rest of his book to describing how culture inflects and shapes technology. This is not surprising given his commitment to technology as narrative. Since the social dimensions of technology are baked right into the definition (as I have argued is the case in Joe's definition), students are quick to see some of the problems with attributing to technology a deterministic autonomy.

In my class, I present technological determinism as one theory among others that that try to explain the relationship between technology and society. I place the kind of determinism Nye discusses on the far end of a continuum of theories that try to explain the same phenomenon. I label the extreme form of determinism that attributes full autonomy to technology and practically none to humanity in explaining the relationship between technology and society "hard technological determinism." This form of determinism suggests that the relationship between technology and society is a unilateral relationship in which technology is the sole force that shapes society (T → S). All social components of the world are to be understood solely as the result of the inexorable will of technology. The extremity of this position invites simple counterexamples. Nye uses many, but the one that usually sticks with students is the story of the Japanese and the gun.

> The gun would appear to be the classic case of a weapon that no society could reject once it had been introduced. Yet the Japanese did just that. They adopted guns from Portuguese traders in 1543, learned how to make them, and gradually gave up the bow and the sword. As early as 1575 guns proved decisive in a major battle (Nagoshino) [*sic*], but then the Japanese abandoned them, for what can only be considered cultural reasons. The guns they produced worked well, but they had little symbolic value to warriors, who preferred traditional weapons. The government restricted gun production, but this alone would not be enough to explain Japan's reversion to swords and arrows. Other governments have attempted to restrict gun ownership and use, often with little success. But the Japanese samurai class rejected the new weapon, and the gun disappeared. It re-entered society only after 1853, when Commodore Perry sailed his warships into Japanese waters and forced the country to open itself to the West. Japan's long, successful rejection of guns is revealing. A society or a group that is able to act without outside interference can abolish a powerful technology. (Nye 2006, 16)

Since hard technological determinism makes no room for society to influence technology, the theory is vulnerable to simple counterexamples. In fact, I often find myself trying to engage in the mental gymnastics required to give hard technological determinist theories a fair hearing in class. This is often accomplished by trying to impress upon students how seriously otherwise very intelligent people used to take this extreme theory. Here, Joe's piece "The Autonomy of Technology" (Pitt 1987) provides a useful counterpoint. While Joe's primary target is Ellul in this piece, his arguments against attributing causal powers to technology apply to all theories of this sort and highlight the ubiquity of theories that offer such arguments. Joe rightly points out that, "in addition to the fact that it is empirically false that technology has these characteristics," the characteristics attributed to technology by hard technological determinism, "[T]he profit in treating technology in this way, to the extent there is any, is only negative. It lies in removing the responsibility from human shoulders for the way in which we make our way around in the world" (Pitt 2000, 87).

Moving away from the extreme position of hard technological determinism, I next introduce students to soft technological determinism.[6] This point in the spectrum of theories describing the relationship between technology and society represents theories that allow for some changes to be made to technology in the early stages of its research, development, and adoption but argues that over time, as the technology gains momentum[7] and becomes widely used, it can achieve a technological lock-in whereby it becomes less malleable and less open to change from societal forces. Here, the relationship between technology and society is understood to be a conditional bi-lateral relationship whereby technology shapes society and society can shape technology prior to technological momentum achieving technological lock-in (T ←→ S). Many students find this position palatable. For example, it meshes well with their understanding of the difficulties of moving away from a fossil fuel-based transportation system. Despite people wanting to switch to cleaner sources of energy, the student might argue, the technological infrastructure prevents this change from taking place in the way (otherwise autonomous) people would want. It also makes sense of the idea that new technologies can provide the solution to social problems (if only enough people buy in). For example, the idea that information and communication technologies have the power to somehow force social change. The most cited example of this is Nicholas Negroponte's declaration, "digital technology can be a natural force drawing people into greater world harmony" (Negroponte 1995, 230). Presumably, once enough people buy in, the rest are forced to follow. This, again, meshed well with many people's experience with having to (to varying degrees of necessity) use social media to communicate with colleagues, friends, and family.

While soft technological determinism does not suffer from some of the simple counterexamples that hard technological determinism suffers from (in fact, it seems to explain the story of the Japanese and gun quite well, given that the gun was eventually adopted), it still places shackles on humanity. In arguing that once technological lock-in is achieved, humanity is powerless against technological force, independent of human will, many students still feel as though the theory goes too far. They point to the story of the Amish (also used in Nye's chapter arguing against determinism) as a counterexample that stands up against soft technological determinism (unlike the Japanese, who eventually did accept the gun into their society [as described above] the Amish continue to successfully guide how modern technologies are introduced into their communities). Doesn't the fact that there exist small communities situated right in the heart of some of the most industrialized, modern societies that have formal structures by which they make decisions about which technologies to allow into their communities mean that all of us have the ability to make the same choice? Doesn't the notion of technological momentum and the resultant technological lock-in really just amount to a claim that it is *really* hard to push back against a widely used technology, not that it is impossible? What students are insisting upon here is the autonomy of humanity.

At this point in the course, I go back to the spectrum of theories describing the relationship between technology and society being introduced and provide students the language to discuss those theories that occupy the end of the spectrum opposite that of hard technological determinism. On this far end of the spectrum I place theories of social constructivism. These I present as the obverse of the hard technological determinist theories. I label this set of theories "all the way down"[8] social constructivist theories to emphasize the unilateral relationship between society and technology (S → T). Technology, under these theories, is to be understood as nothing more than the reification of social values. There is no element of technology, no matter how far down one explores the technological artifice in question, that shapes society, no causal role for artifact alone to affect society. Rather, it is social choices made "all the way down." Technology is an epiphenomenon of social and political interests. The technologies of automation and mechanization did not shape the social structure of the industrial revolution. They did not bring about the political empowerment allowed by the unequal concentration of wealth in the hands of the few. They did not make permanent underclasses of workers and empower the owners of production. Rather, they are themselves the expression of these preexisting class differences.

Human beings being viewed as dehumanized cogs in a machine is, according to all the way down social constructivism, not the result of technology, as so many have argued in the past, but the very cause of the introduction of technologies of mechanization and automation. These technologies are the reification of the dehumanization of entire classes of humans. This strong, all the way down, brand of social constructivism informed much of early STS scholarship.

As with hard technological determinism, this position suffers from an absolutism that is difficult to defend. It does not account for most of our lived experiences in which we can feel shaped by technology. Nor does it account for how we usually understand the history of humanity's relationship with technology. While it allows us to shake teleological histories of the inevitable unfolding of society according to the inexorable will of technology, it fails to account for the ways in which historical choices regarding technology affect what choices are available for us to make today. It also creates a serious problem of anti-realism. If social causes explain technology all the way down, where do the objects of science come in? Where does the world come in? How can we explain the reliability and effectiveness of many technologies if we claim that science, too, is socially constructed all the way down? These are problems that many STS scholars are reexamining today in very public ways.[9]

To address these concerns, I introduce the last point on the spectrum of theories describing the relationship between technology and society. For symmetry's sake and for ease of conceptual categorization for students, I next introduce the counterpart to soft technological determinism. This point I label "co-productionism." It is, I tell students, and submit to readers here, the most widely held theory of the relationship between society and technology in science and technology studies today. It is simply the claim that the social affects the technical and the technical affects the social in complicated, but uncoverable and understandable, ways (S ← → T). To put it more precisely,

> Coproduction is shorthand for the proposition that the ways in which we know and represent the world (both nature and society) are inseparable from the ways in which we chose to live in it. Knowledge and its material embodiments are at once products of social work and constitutive of forms of social life; society cannot function without knowledge any more than knowledge can exist without appropriate social supports. Scientific knowledge, in particular, is not a transcendent mirror of reality. It both embeds and is embedded in social practices, identities, norms, conventions, discourse, instruments, and institutions—in short, in all building blocks of what we term the social. The same can be said even more forcefully of technology. (Jasanoff 2004, 2–3)

This node on the spectrum tries to be the goldilocks between the extremes of technological determinism and social constructivism.

A useful metaphor in my classes is the evolutionary relationship between humanity and technology. Humanity and technology are inseparable in evolutionary terms. As the human brain grew larger and more complex, we were able to learn how to manipulate our environment in more complex ways. This ability to manipulate the environment with more complexity pushed our brains to become itself more complex in ways that allowed us to find increasingly complex ways to manipulate the environment.

Technology is not autonomous in this story. Technology is indeed humanity at work, but so too is humanity technology at work, but humanity is also technology at work. Humans in this story are not fully autonomous, either. But this is not a remarkable point. No one argues that human beings are absolutely autonomous in the sense that everything they do is the sole result of an unfettered, unshaped free will. Being born into today's technological society can feel a lot like having been born into a maze or a labyrinth. Your options and choices are bound by the technological apparatus. Technologies make things that were impossible possible and things that were possible impossible. They make plenty of room for new choices to be made, choices that are limited and shaped by those that were made prior.

The question of how our choices are shaped or limited by technology and how technology itself can be influenced by social choices is an empirical one. The aim of contemporary STS is to tell the complicated story about how technology has come to shape our lives in the way it has, how it continues to do so, and what pathways are available in the future given the social choices available. These social choices are choices about the good life. These are normative choices. These are the very things we refer to in our definition of technology. Nye's "intentions" and Pitt's "goals" are components of our definition of technology that deal with these questions (among others) directly.

The ways in which the shaping of choices takes place is described fantastically in one of Joe's pieces (Pitt 2008) I use regularly in my computer science course: "The Social and Ethical Dimensions of Information and Communication Technologies." This piece, especially when supplemented with a bit of Sherry Turkle (2017), shows how the choices we make about new technologies can shape how other choices are made (in ways that might sometimes elicit disapprobation). In this piece, Joe laments the ubiquity of the iPod and the resultant lack of heated debate in the graduate student halls. What's wrong with this? Nothing, Joe says. "The iPod itself is a piece of technology. As a piece of technology, it is neither right or wrong, good or bad" (Pitt 2008, pp. 161–165). He goes on to insist that it is the use to which a technology is put that gives us the context to say a technology is good or bad. But this use is already understood as an essential element of our definition of technology. Now, I believe we have come to a semantic point. I want to say that the iPod as an artifact is neither good nor bad. But, the moment we understand the iPod as a technology we are already discussing the use to which it is being put. The only reason left to be worried about the idea that the iPod itself might be good or bad has to do with our reluctance to ascribe autonomy to iPods.

The spectrum of theories I have laid out here shows how autonomy (of both humanity and technology) is distributed across a variety of theoretical approaches to understanding the relationship between technology and society. What I hope to have made clear is that there are approaches to understanding this relationship that account for the autonomy of humanity and the ethical evaluation of technologies themselves while avoiding the charges of both technological and social determinism.

I want engineers to make design decisions that are consciously informed by the social and ethical choices that are being made during the design process. To achieve this, I teach engineers-in-the-making (engineering students) how to go about gathering information so that they can make such informed decisions. The first step in this pedagogical pursuit is the defining technology in such a way that making *choices* about the *social and ethical* components of a technology in the process of research and design is conceptually possible. This is compatible with the coproduction account of technology and society that allows. This is the *use* to which I am putting my definition of technology. When I present the variety of ways to define technology and how they interact with the spectrum of theories describing the relationship between society and technology, I am conscious of which of these theories best serves the goal of teaching engineers-in-the-making how to create technologies in a way that consciously pursues socially just outcomes. I am making choices about the social and ethical components of my pedagogical approach to shaping engineers-in-the-making. As Pitt puts it, in his piece, "Human Beings as Technological Artifacts," "We live in a technological world that at least appears to be wrapping us up in electronics and other technologies without asking for our consent. The ability to select those technologies I want to be associated with is important to who I am and who I will become" (Pitt 2006, 133). I borrow this concept for the title of this piece so as to bring attention to the kinds of choices engineers-in-the-making need to make in deciding what kind of engineer they will become. Through the philosophical investigation of the nature of technology, I hope to make visible to them the ways in which ethical considerations can and do enter into the processes involved in developing new technology.

A while back I referred to Joe's definition of technology as "infamous." As social scientists-in-the-making, the graduate students in STS at Virginia Tech were happy to hear technology defined as "humanity at work." If the social sciences are your hammer, this is a great way of turning technical fields into a nail. However, his insistence that there is nothing normative about technologies in and of themselves turned off a lot of the STS students. What I hope to have shown is that there is a path through the extreme variants of theories employed by STSers of the past, as represented on my spectrum, which can allow for Joe and the current batch of STS students to play nice.

NOTES

1. It was only after I had written this introduction that I realized how closely these sentiments mirror Joe's own in the introduction to his article, "Against the Perennial" in which he describes a similar process of coming to realize he had been "seeking the wrong grail" in his philosophical pursuits.
2. See Propositions 6.54 and 7.
3. A perfectly reasonable requirement from the people signing my paychecks.
4. I would like to acknowledge that Dr. Govind Gopakumar was the primary author of this inherited syllabus.
5. Apart from Nye's insistence that recent fieldwork in primatology has demonstrated that nonhuman animals use tools in a way that allows them to be understood to be said to be engaging with technology, hence my parenthetical "human or otherwise" here, this definition tracks very well with Joe's. The notion that human beings are the only kind of things that can form goals, intentions, and desires, and thereby make use of artifacts as tools/technologies, has been effectively shown to be too exclusionary. Also, I want my definition of technology to be open enough to cover the use of technology by super intelligent aliens. I quickly pass over it here because it requires a simple fix and I'm sure Joe's love of science fiction would lead him to agree with me on the aliens thing at least. Suggestion: from "Technology is humanity at work" to something like, "Technology is 'things like humans' at work?"
6. I want to note that at this point I am getting far away from anything Nye says in *Technology Matters*.
7. This borrows from Thomas P. Hughes's work. See Hughes (1989, 45).

8. Named so after the "turtles all the way down" expression of the infinite regress problem. The idea is that, no matter how far down you go in the explanation of the relationship between technology and society, you will always encounter a social cause.

9. See Latour's recent piece in NYT. The causes of the "science wars" of the 1990s reside in these issues as well. The same conversations that shaped these "wars" remain ongoing. Though, I show in the next section that most of STS has moved beyond this brand of absolutist social determinism.

REFERENCES

Hughes, Thomas. 1989. "The Evolution of Large Technological Systems" in *The Social Construction of Technological Systems: New Directions in the Sociology and History of Technology*, edited by Wiebe E. Bijker, Thomas Parke Hughes, and Trevor J. Pinch. Cambridge: MIT Press.

Jasanoff, Sheila (Editor). 2004. *States of Knowledge: The Co-Production of Science and the Social Order*. New York: Routledge.

Keulartz, Jozef, et al. 2002. *Pragmatist Ethics for a Technological Culture*. Dordrecht: Kluwer.

Latour, Bruno. https://www.nytimes.com/2018/10/25/magazine/bruno-latour-post-truth-philosopher-science.html.

Negroponte, Nicholas. 1995. *Being Digital*. New York: Vintage.

Nye, David. 2006. *Technology Matters: Questions to Live With*. Cambridge: MIT Press.

Pitt, Joseph C. 1987. "The Autonomy of Technology" in *Technology and Responsibility*, edited by Paul Durbin, pages 99–114. London and New York: Springer.

Pitt, Joseph C. 2000. *Thinking about Technology: Foundations of the Philosophy of Technology*. Seven Bridges Press.

Pitt, Joseph C. 2003. "Against the Perennial: Small Steps toward a Heraclitian Philosophy of Science." *Techné*. 7 (2): 57–65.

Pitt, Joseph C. 2004. "Pragmatist Ethics in the Technological Age." *Techné*. 7 (3): 32–38.

Pitt, Joseph C. 2006. "Human Beings as Technological Artifacts" in *Defining Technological Literacy: Towards an Epistemological Framework*, edited by John Dakers. London and New York: Springer.

Pitt, Joseph C. 2008. "Don't Talk to Me" in *iPod and Philosophy: iCon of an ePoch*, edited Dylan Wittkower. Chicago: Open Court Press.

Turkle, Sherry. 2017. *Alone Together: Why We Expect More from Technology and Less from Each Other*. Hachette, UK.

Wittgenstein, Ludwig. 2013. *Tractatus logico-philosophicus*. New York: Routledge.

Chapter 10

Gravity and Technology

Allan Franklin

Joe Pitt is well known for his argument that technology is essential for both the practice and progress of science. In looking at two recent and very important experimental results, the discovery of the Higgs boson by the CMS and ATLAS collaborations at the Large Hadron Collider and the observation of gravity waves by the LIGO-Virgo collaboration, the experimental apparatuses were technological tours de force. Without advanced technology, those experiments would not have been possible. The question of which comes first between science or technology has no simple, universal answer. Sometimes the advances in technology precede the advances of science, in other instances the science came first. As we shall see, advances in technology have improved the precision in measurements of *G*, the universal gravitational constant in Newton's Law of Universal Gravitation. Unfortunately, this precision has not led to agreement. The later measurements demonstrated the importance of small confounding effects. There are also instances in which new technology is not needed. Sometimes a measurement may be "good enough" for the intended purpose. Thomas Young's experiment on the interference of light, which was of great importance in establishing the wave nature of light, did not require new technology.

In examining questions concerning the use of complex, technologically advanced apparatus in experiments there are several important issues, some of which I have discussed at length elsewhere. These include the varying roles of experiment and the question of how one calibrates an experimental apparatus. Several of these discussions appeared in *Perspectives on Science*, then edited by Joe Pitt (Franklin 1993a, 1993b, 1995, 1997). I am grateful to Joe for his support and encouragement when I was shifting my research from experimental particle physics to history and philosophy of science.

In this chapter I will examine several episodes from the history of gravity measurements to help illuminate the connection between science and technology. We will find that, in some cases, advances in technology have greatly improved the measurements, whereas in others, the progress is not so clear.

FALLING BODIES

In this section I will discuss the history of experiments on falling bodies, from Galileo's mythical experiments at the Leaning Tower of Pisa to modern experiments on the Universality of Free Fall. Galileo's experiments involved essentially no technology, whereas in the following history we see increasingly advanced technology used in such experiments from Newton's use of the simple pendulum, through Eötvös's use of a torsion balance, and finally to the use of a torsion pendulum by the Eöt-Wash collaboration. This history involves the use of more advanced technology to provide both more precise measurements and increasingly stringent tests of theory. We will also see that the theories under test varied from Aristotle's theory of falling bodies to Newton's Law of Universal Gravitation, to Einstein's General Theory of Relativity.

Galileo, the Leaning Tower of Pisa

I will begin with an account of Galileo's mythical experiment at the Leaning Tower of Pisa. In that experiment, Galileo is said to have dropped two unequal weights from the top of the tower and observed with his own eyes that they fell at the same rate, striking the ground almost simultaneously. Thus, the experiment refuted Aristotle's view that bodies fall with speeds proportional to their weight. As Lane Cooper (1935) has convincingly argued, there is serious doubt that Galileo ever performed the experiment. Galileo did not discuss any such experiment at the Leaning Tower in any of his writings. Nevertheless, as recounted by Viviani, one of Galileo's students, twelve years after Galileo's death, Galileo performed the experiment several times before the assembled students and faculty at the University of Pisa around 1590.

> And then, to the dismay of all the philosophers, very many conclusions of Aristotle were by him [Galileo] proved false through experiments and solid demonstrations and discourses, conclusions which up until then had been held for absolutely clear and indubitable; as, among others, that the velocity of moving bodies of the same material, of unequal weight, moving through the same medium, did *not* mutually preserve the proportion of their weight as taught by Aristotle, but all moved at the same speed; demonstrating this with repeated

> experiments from the top of the Campanile of Pisa in the presence of the other teachers and philosophers and the whole assembly of students. (Viviani, quoted in Cooper 1935, 26)

I note that even for a direct observation Viviani cites the presence of other observers to increase the credibility of the observation.

Galileo did discuss the phenomenon of falling bodies in his *Discourse on Two New Sciences* (1974). In that dialogue, Salviati, Galileo's surrogate, states,

> Aristotle says: "An hundred-pound iron ball, falling from the height of 100 braccia [approximately 225 feet], hits the ground before one of just one pound ball has descended a single braccio."[1] I say that they arrive at the same time. You find, on making the experiment, that the larger anticipates the smaller by two inches; that is when the larger one has strikes the ground, the other is two inches behind it. And now you want to hide, behind those two inches, the ninety-nine braccia of Aristotle, and speaking only of my tiny error, remain silent about his enormous one. (Galileo 1974, 68)

Sagredo, a neutral observer, discusses a slightly different experiment. "But I, Simplicio, who have made the test, assure you that a cannonball that weighs 100 pounds (or two hundred or even more) does not anticipate by even one span the arrival on the ground of a musket ball of no more than half [an ounce], both coming from a height of two hundred braccia" (Galileo 1974, 66). Notice that in neither case is there any mention of the Leaning Tower. Galileo is also discussing two balls, of the same material.[2] Galileo is arguing that even though he has observed a small difference, the experiment is "good enough" to refute the Aristotelian theory of falling bodies.

The Eötvös Experiment

In this section we will see that technological advancements in the experimental apparatuses used allowed more stringent tests of the equality of free fall for objects of different mass and/or different composition. Newton used a simple pendulum, which was certainly an improvement over Galileo's direct observation. Eötvös made use of a torsion pendulum, which had been used in the late eighteenth century by Henry Cavendish to measure the density of the Earth, and by Charles-Augustin de Coulomb, to find the law of force between electrical charges (see discussion below). Eötvös's apparatus was considerably more advanced than that of Coulomb. Further improvements were made by the Eöt-Wash group.

The 1687 publication of Newton's *Principia* provided a theoretical basis for the equality of free fall for objects of different materials. This depended on the hypothesis that the ratio of the gravitational mass of an object to its inertial mass was constant for all substances. The gravitational mass is the mass that gives rise to a gravitational force when an object is placed in a gravitational field caused by a second gravitational mass. Thus, $F_G = m_G \times g$, where g is the gravitational field at the surface of the Earth.[3] The inertial mass is the resistance of an object to changes in its state of motion. Thus, $F = m_I \times a$. Applying Newton's second law to a freely falling object at the surface of the Earth we find $m_G \times g = m_I \times a$. Then $a = (m_G/m_I) \times g$. If the ratio (m_G/m_I) is constant for all substances, then all objects will fall at the same rate.

This had presumably been shown by Galileo and others. Newton improved this measurement using two pendulums "Others have long since observed that the falling of all heavy bodies toward the earth (at least on making an adjustment for the inequality that arises from the very slight resistance of the air) takes place in equal times, and it is possible to discern that equality of the times, to a very high degree of accuracy, by using pendulums. I have tested this with gold, silver, lead, glass, sand, common salt, wood, water and wheat" (Newton 1999, pp. 806–7). The period of the pendulums is proportional to $(m_G/m_I)^{1/2}$ so if the ratio is constant the periods will be equal. Newton showed that the ratio was constant to one part in 1,000. In 1832 Bessel, using a similar technique, lowered this limit to one part in 60,000.

In the late nineteenth century Roland von Eötvös and his collaborators began a series of experiments that measured the equality of fall for different substances, thus the Universality of Free Fall (UFF).[4] In 1890 Eötvös lowered that limit to one part in 20,000,000 (Eötvös 1890). The data for his most famous experiment was taken in the period 1904–09 but not published until 1922 (Eötvös, Pekar et al. 1922).

The apparatus for the Eötvös experiment is shown schematically in figure 10.1. Eötvös used a torsion balance, a considerable technological improvement, to make his measurements.[5] As shown in figure 10.1, the gravitational force is not parallel to the fiber due to the rotation of the Earth. If the gravitational force on one mass differs from that on the other, the rod will rotate about the fiber axis. Reversing the masses should give a rotation in the opposite direction. Eötvös's original apparatus typically contained a standard platinum mass and contained a comparison mass, suspended by a copper-bronze fiber. Because the experiment was originally designed to measure vertical gravity gradients, the two masses were suspended at different heights, which made the apparatus significantly more sensitive to such gradients than it was to anomalous gravitational accelerations. To calculate and subtract the effect of these gradients Eötvös performed four measurements: with the torsion bar oriented in the North–South, the South–North, the East–West, and

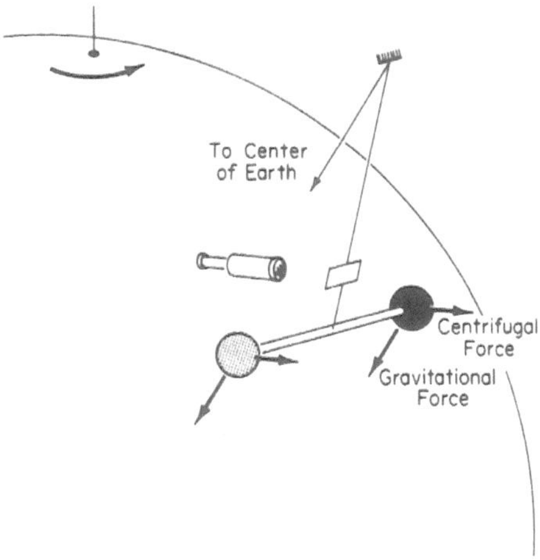

Figure 10.1 A Schematic View of the Eötvös Experiment. *Source*: From *Was Einstein Right? Putting General Relativity to the Test* by Clifford M. Will, copyright © 1986. Reprinted by permission of Basic Books, an imprint of Hachette Book Group, Inc.

the West–East directions, respectively. (For a detailed discussion of both the apparatus and the methods of data analysis, see Fischbach, Sudarsky et al. 1988, 8–12.)

The results of Eötvös and his collaborators for the difference in the acceleration of falling bodies of different substances gave an average value for $x = \Delta a/g = (-0.002 \pm 0.001) \times 10^{-6}$, where Δa is the difference in acceleration of the different bodies and g is the acceleration due to gravity at the earth's surface. This was equal to approximately a part in a billion. The individual values for x, relative to platinum as the standard were (Magnalium, +0.004 ± 0.001; Snakewood, −0.001 ± 0.002; Copper, +0.004 ± 0.002, Water, −0.006 ± 0.003; Crystalline cupric sulfate, −0.001 ± 0.003; Solution of cupric sulfate, −0.003 ± 0.003; Asbestos, +0.001 ± 0.003; Tallow, −0.004 ± 0.003). They concluded, "We believe we have the right to state that x relating to the Earth's attraction does not reach the value of 0.005×10^{-6} for any of these bodies" (Eötvös et al. 1922, 164).

The Eöt-Wash Experiments, the Fifth Force, and the Weak Equivalence Principle

In 1986 Ephraim Fischbach, Sam Aronson, Carrick Talmadge, and their collaborators proposed a modification of Newton's Law of Universal Gravitation (Fischbach, Aronson et al. 1986).[6] The modification took the form $V = (-Gm_1m_2/r)[1 + \alpha e^{-r/\lambda}]$. The first term is the Newtonian gravitational potential. The second term is the Fifth Force potential, where α is the strength of the Fifth Force and λ is its range.[7] The Fifth Force was hypothesized to be composition dependent.[8] The Fifth Force between two lead objects would differ from the Fifth Force between a lead object and a copper object.

The proposal was based on three tantalizing pieces of evidence: (1) the difference between measurements of G, the Universal Gravitation constant in Newton's law, in mineshafts and in the laboratory; (2) an observed energy dependence in the CP violating parameters in K^0 meson decay; and (3) a reanalysis of the Eötvös experiment.

As discussed in the previous section, Eötvös and his collaborators had set a limit on the violation of the Universality of Free Fall of approximately one part in a billion. Fischbach and his collaborators reanalyzed Eötvös's data. They plotted the data reported by Eötvös for Δk, the fractional difference in acceleration, as a function of $\Delta(B/\mu)$, the difference between the ratio of baryon number to mass in units of atomic hydrogen mass for different substances. They found a startling composition-dependent effect (figure 10.2).

In early 1987 the first two experimental tests of the Fifth Force hypothesis were reported. The Eöt-Wash collaboration presented strong evidence against the existence of the Fifth Force (Stubbs, Adelberger et al. 1987).[9] Their experimental apparatus is shown in figure 10.3. It, too, was designed to search for a substance-dependent, intermediate range force, and was located on a hillside on the campus of the University of Washington in Seattle.[10] If

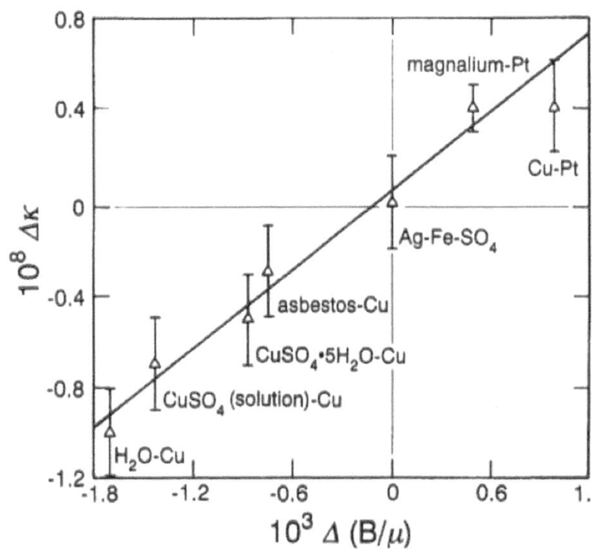

Figure 10.2 Plot of Δk as a Function of $\Delta(B/\mu)$. *Source*: Reprinted figure with permission from Fischbach, E., S. H. Aronson, et al., *Physical Review Letters*, 56: 4, 1986. Copyright 1986 by the American Physical Society.

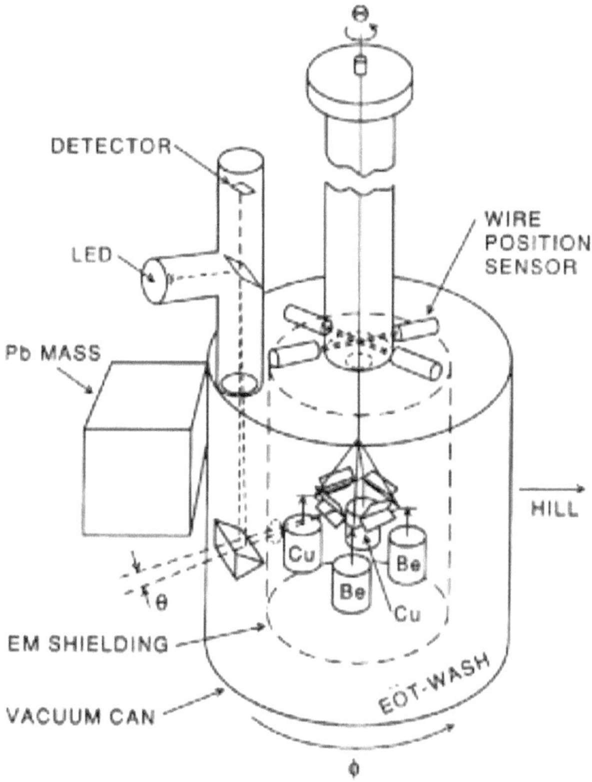

Figure10.3 Schematic View of the Eöt-Wash Torsion Pendulum Experiment. *Source*: Reprinted figure with permission from Stubbs, C. W., E. G. Adelberger, et al., *Physical Review Letters*, 58: 1070, 1987. Copyright 1987 by the American Physical Society.

the hill attracted the copper and beryllium test bodies differently, the rotating torsion pendulum[11] would experience a net torque. The pendulum, or baryon dipole, could be rotated with respect to the outside can in multiples of 90° and the entire system rotated slowly ($T_{can} \approx 6 \times 10^3$ s). If there were a differential force on the copper and beryllium, one would expect to find a torque that varied with θ, the angle of the can, with respect to some fixed geographical point. They detected torques by measuring shifts in the equilibrium angle of the torsion pendulum. Great care was paid to systematic effects that might either produce a spurious signal or cancel a real signal. To minimize asymmetries the test bodies were machined to be identical within very small tolerances. Electrostatic forces were minimized by coating both the test bodies and the frame with gold, and by surrounding the torsion pendulum with a grounded copper shield. Magnetic shielding was also provided and Helmholtz coils reduced the ambient magnetic field to 10 mG. Reversing the current in the

Helmholtz coils caused a_1, the signature of the interaction, to change by 3.8 ± 2.3 μrad. Scaling that result to their normal operating conditions implied a systematic error at the level of 0.1 μrad. Gravity gradients, which might result in a spurious signal if all the test bodies were not in a plane, were reduced by placing an 80 kg lead mass near the apparatus. They set an upper limit of 0.19 μrad on any possible spurious signal due to such gradients. The most serious source of possible error was due to the "tilt" of the apparatus, which was very sensitive to such tilts. A deliberately induced tilt of 250 μrad produced a spurious a_1 signal of 20 μrad. They measured the tilt sensitivity of their apparatus carefully and corrected their data for any residual tilt. In addition, they included in their final results only those data for which the tilt was less than 25 μrad and for which the tilt correction to a_1 was less than 0.71 μrad. This was the only cut made on their data, although they noted that including all of the data gave results in good agreement with their selected sample. They also determined an upper limit of 0.11 ± 0.19 μrad due to thermal effects.

The Eöt-Wash results are presented in figure 10.4 (Stubbs, Adelberger et al. 1987).[12] There is no apparent signal, although there is an offset of 4 μrad. The theoretical curves were calculated using values of α and λ of 0.001 and 100 m, respectively. The published version of a paper presented at the Moriond workshop (Raab, Adelberger et al. 1987) contains theoretical curves calculated with $\alpha = 0.01$, and shows far more disagreement with the data (see figure 10.5).[13] The value used by Fischbach at the time was 0.007. The paper in *Physical Review Letters* (PRL) understated the extent of the disagreement between the Fifth Force theory and the Washington results. The Washington group concluded: "Our results rule out a *unified* [emphasis added] explanation of the apparent geophysical and Eötvös anomalies in terms of a new baryonic interaction with $10 < \lambda < 1400$ m and make it highly improbable that the systematic effects in the Eötvös data are due to a new fundamental interaction coupling to B [the baryon number]" (Stubbs, Adelberger et al. 1987, 1072).

It is important to note here that the Eöt-Wash collaboration not only constructed a new and more sensitive torsion pendulum, but their experimental procedures minimized possible confounding effects or demonstrated that they were too small to affect their result. As they later remarked, "Our differential accelerometer—a highly symmetric rotating torsion balance—incorporated several innovations that effectively suppressed systematic errors. All known sources of systematic error were demonstrated to be negligible in comparison to our fluctuating errors" (Adelberger, Stubbs et al. 1990, 3267).

The Eöt-Wash group would continue their measurements into the twenty-first century, making improvements and setting ever more stringent limits on the Universality of Free Fall. In 1990 they stated, "In terms of the weak equivalence principle in the field of the Earth, our 1σ result corresponds to $m_i/m_g(\mathrm{Cu}) - m_i/m_g(\mathrm{Be}) = (0.2 \pm 1.0) \times 10^{-11}$" (Adelberger, Stubbs et al. 1990,

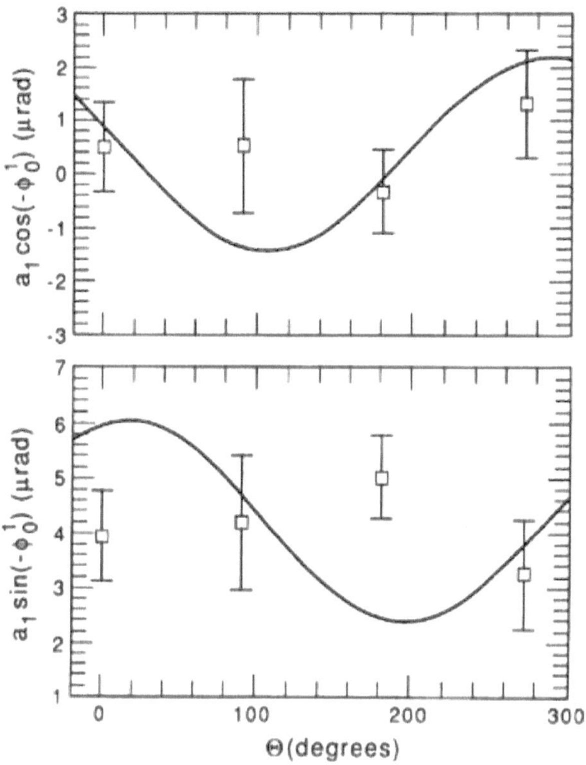

Figure 10.4 Deflection Signal as a Function of θ, the Variable Angle of the Stage. The theoretical curves correspond to the signal expected for $\alpha = 0.001$ and $\lambda = 100$ m. *Source*: Reprinted figure with permission from Stubbs, C. W., E. G. Adelberger, et al., *Physical Review Letters*, 58: 1071, 1987. Copyright 1987 by the American Physical Society.

3267). In 1994 even lower limits were set. The Eöt-Wash collaboration used the Eötvös ratio, $\eta = 2\,((m_g/m_i)_1 - (m_g/m_i)_2)/((m_g/m_i)_1 + (m_g/m_i)_2)$ as a measure of their test of the Universality of Free Fall.[14] "In terms of the classic UFF parameter η, our Earth-source results are η(Be,Cu) = $(-3.0 \pm 3.6) \times 10^{-12}$ and η(Be,Al) = $(-0.2 \pm 2.8) \times 10^{-12}$ where all our errors are 1σ. Thus our limit for UFF violation for Be and a composite Al/Cu body is $\eta = (-1.1 \pm 1.9) \times 10^{-12}$" (Su, Heckel et al. 1994, 3614).

The Eöt-Wash group later reported a new result using an interesting variant on their previous experimental apparatus (Gundlach, Smith et al. 1997). In their previous work, the group had used a torsion balance mounted on a rotating platform to measure the differential acceleration of various substances toward a local hillside, and to other sources such as the Sun, the Earth, and the galaxy. In their latest experiment the experimenters used a rotating three-ton ^{238}U attractor to measure the differential acceleration of lead and copper

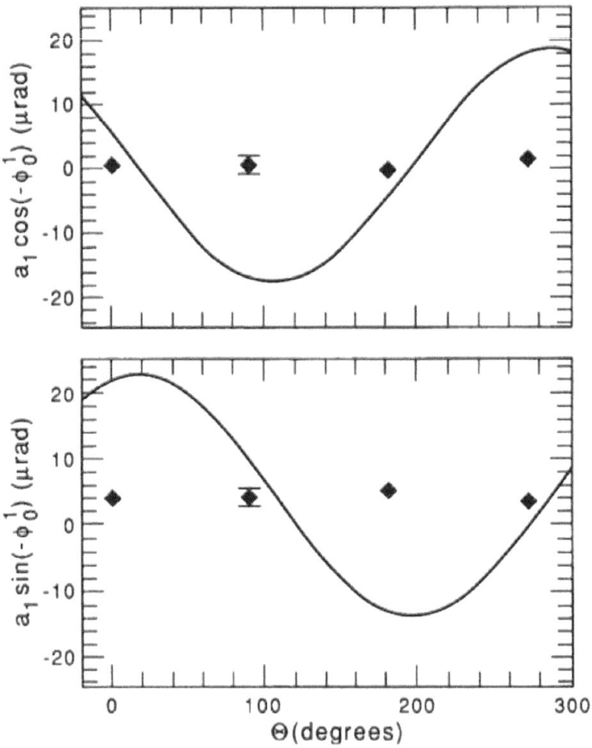

Figure 10.5 Deflection Signal as a Function of θ, the Variable Angle of the Stage. The theoretical curves correspond to the signal expected for $\alpha = 0.01$ and $\lambda = 100$ m. *Source*: Raab, F. J. (1987). "Search for an Intermediate-Range Interaction: Results of the Eot-Wash I Experiment." New and Exotic Phenomena: Seventh Moriond Workshop. O. Fackler and J. Tran Than Van. Gif sur Yvette, Editions Frontieres.

masses placed on a torsion balance. The Röt-Wash[15] apparatus is shown in figure 10.6. The surroundings of the torsion balance were temperature controlled to guard against possible temperature effects. The ^{238}U was counterbalanced by 820 kg of Pb so the floor would not tilt as the attractor revolved. As discussed earlier, tilt was a significant source of possible background effects in the Eöt-Wash experiments. The reason for the modification of the apparatus was that their previous experiment had been unable to test for forces with a range from 10 to 1,000 km. The new apparatus, using a local source, allowed such a test. The experimenters concluded "We found that $\alpha_{Cu} - \alpha_{Pb} = (0.7 \pm 5.7) \times 10^{-13}$ cm/s^2 . . ." (Gundlach, Smith et al. 1997, 2523).

The Eöt-Wash collaboration continued their extensive study of the equivalence principle with a new and improved torsion balance (Schlamminger, Choi et al. 2008). Their results for the difference in acceleration for beryllium and titanium test masses, in the northern and western directions, are shown in figure 10.7. A violation of the equivalence principle would appear as a

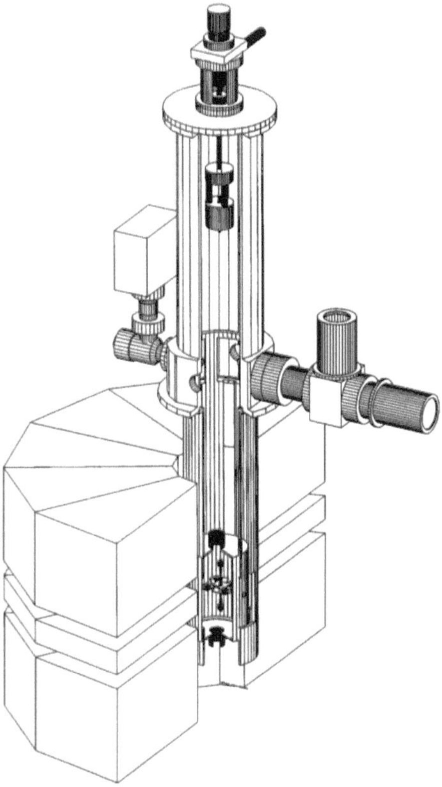

Figure 10.6 Schematic View of the Röt-Wash Instrument. *Source*: Reprinted figure with permission from Gundlach, J. H., G. L. Smith, et al., *Physical Review Letters*, 78: 2524, 1997. Copyright 1997 by the American Physical Society.

difference in the means of the runs taken with the masses in different orientations. The small offset was due to a systematic error, which did not affect their conclusion. "We used a continuously rotating torsion balance instrument to measure the acceleration difference of beryllium and titanium test bodies toward sources at a variety of distances. Our result $\Delta\alpha_{N,Be\text{–}Ti} = (0.6 \pm 3.1) \times 10^{-15}$ m/s^2 improves limits on equivalence-principle violations with ranges from 1m to ∞ by an order of magnitude. The Eötvös parameter is $\eta_{Earth,\,Be\text{–}Ti} = (0.3 \pm 1.8) \times 10^{-13}$" (Schlamminger, Choi et al. 2008, 041101-1). Recall that their previous best limit for η was 1.1×10^{-12}.

Discussion

This episode is almost a paradigmatic case in which improving technology provided more accurate and more precise experimental results and more

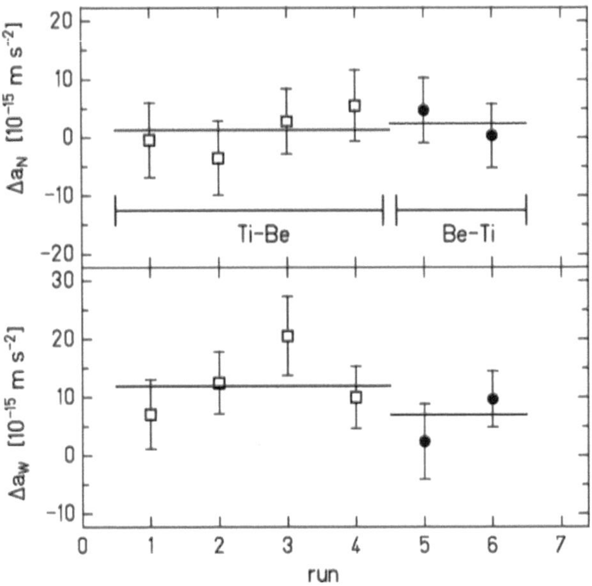

Figure 10.7 Shown are measured differential accelerations toward north (*top*) and west. After the first four data runs, the Be and Ti test bodies were interchanged on the pendulum frame. A violation of the equivalence principle would appear as a difference in the means (*lines*) of the two data sets. The offset acceleration is due to systematic effects that follow the pendulum frame but not the composition dipole. *Source*: Reprinted figure with permission from Schlamminger, S., K. Y. Choi, et al., *Physical Review Letters*, 100: 041101-3, 2008. Copyright 2008 by the American Physical Society.

stringent tests of theory. Galileo observed a hand's breadth difference in a fall of 225 feet. Assuming a hand's breadth of approximately six inches we find a fractional difference in fall of about 2×10^{-3}. In comparison, the Eötvös parameter for the 2008 Eöt-Wash experiment is approximately one part in 10^{13}. The precision of the measurements has improved by a factor of 10^{10}, supporting the Universality of Free Fall and the Weak Equivalence Principle and arguing strongly against the presence of a Fifth Force.

HENRY CAVENDISH, THE DENSITY OF THE EARTH, AND *G*, THE UNIVERSAL GRAVITATIONAL CONSTANT

In this section, I will discuss advancing technology that allows for more precise measurements of the density of the Earth and of *G*. Unfortunately, as we shall see, increasing precision does not guarantee greater agreement. Small confounding effects become more important.

Cavendish and the Density of the Earth

Contrary to the views expressed in many introductory physics textbooks, Henry Cavendish did not measure *G*, the gravitational constant contained in Newton's Law of Universal Gravitation. As the title of his paper states, Cavendish conducted "Experiments to Determine the Density of the Earth" (1798). One can use that measurement to determine *G*, but that was not Cavendish's intent. In fact, the determination of *G* was not done until the latter part of the nineteenth century.

Newton had stated that the gravitational force between two objects was proportional to the product of their masses and inversely as the square of the distance between their centers, but he did not explicitly use the universal gravitational constant, *G*.[16] "Gravity exists in all bodies and is proportional to the quantity of matter in each. . . . If two globes gravitate toward each other, and their matter is homogeneous on all sides in regions that are equally distant from their centers, then the weight of either globe toward the other will be inversely as the square of the distance between the centers" (Newton 1999, 810–11).[17]

Cavendish's method was to measure the gravitational force between two masses using a torsion balance. The apparatus is shown schematically in figure 10.8. Briefly, the gravitational force between the small sphere and

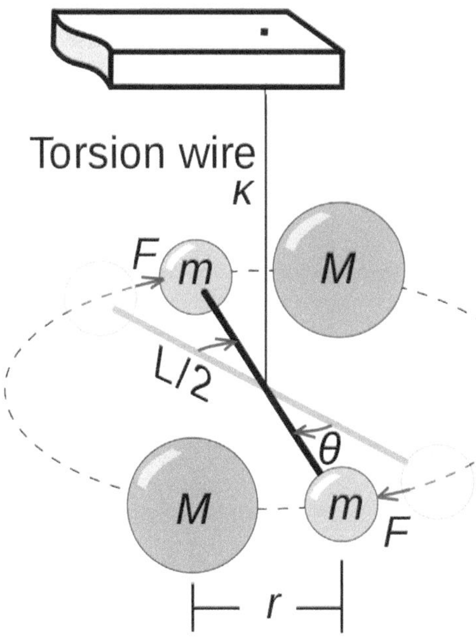

Figure 10.8 Schematic Diagram of the Cavendish Experiment. *Source*: https://en.wikipedia.org/wiki/Cavendish_experiment

the large sphere is measured by measuring the twist of the wire suspending the beam holding the two small masses. To determine that force Cavendish needed to know the torsion constant of the wire. This was found by measuring the period of oscillation of the beam.[18]

Cavendish was extremely careful in avoiding possible confounding effects. He regarded possible temperature gradients in his apparatus as the most important of these. He had found that

> the disturbing force most difficult to guard against, is that arising from the variations of heat and cold; for if one side of the case is warmer than the other, the air in contact with it will be rarefied, and, in consequence, will ascend, while that on the other side, will descend, and produce a current which will draw the arm sensibly side. As I was convinced of the necessity of guarding against this source of error, I resolved to place the apparatus in a room which should remain constantly shut, and to observe the motion of the arm from without, by means of a telescope; and to suspend the leaden weights in such manner, that I could move them without entering the room. (Cavendish 1798, 470–71)

Cavendish also worried about other possible confounding effects such as air resistance, magnetic effects, gravitational effects between parts of his apparatus. He took steps to either minimize these effects or calculate the corrections due to them.

To get the density of the Earth one compares the gravitational force, *F*, between the spheres to *W*, the weight of the small ball. We assume, as Cavendish did, Newton's Law of Universal Gravitation. In modern notation

$$F/W = \left[GmM_{\text{Large}}/r^2\right] / \left[GmM_{\text{Earth}}/R^2\right],$$

where *r* is the distance between the centers of the small and large spheres and *R* is the radius of the Earth.

$$\begin{aligned} F/W &= \left[M_{\text{Large}}/r^2\right] / \left[M_{\text{Earth}}/R^2\right] = \left[M_{\text{Large}}/r^3\right] / \left[M_{\text{Earth}}/R^3\right] \times r/R \\ &= \left[M_{\text{Large}}/r^3\right] / \left[\text{Density}_{\text{Earth}}\right] \times r/R \\ &= \text{Density}_{\text{Earth}} = \left[M_{\text{Large}}/r^3\right] \times r/R \times W/F \end{aligned}$$

All of the terms on the right side of the equation are either measured or known so that one can calculate the density of the Earth.

Cavendish conducted twenty-nine experiments, six with a less stiff wire and twenty-three with a stiffer wire. There is an oddity in Cavendish's final result. Cavendish claimed that the average value he obtained from the first six measurements of the density, 5.48, those with a less stiff wire, was equal to that of the last twenty-three measurements, those found with a stiffer wire. He reported that both sets of measurements gave a density of 5.48[19] and that his final result was 5.48 (figure 10.9). This is not correct. The average of the first six measurements is 5.31 ± 0.22, whereas the average for the last twenty-three

The following Table contains the Result of the Experiments.

Exper.	Mot. weight	Mot. arm	Do. corr.	Time vib	Do. corr.	Density.
1	m. to +	14,32	13,42	′ ″	-	5,5
	+ to m.	14,1	13,17	14,55	-	5,61
2	m. to +	15,87	14,69	-	-	4,88
	+ to m.	15,45	14,14	14,42	-	5,07
3	+ to m.	15,22	13,56	14,39	-	5,26
	m. to +	14,5	13,28	14,54	-	5,55
4	m. to +	3,1	2,95		6,54	5,36
	+ to —	6,18	-	7,1	-	5,29
	— to +	5,92	-	7,3	-	5,58
5	+ to —	5,9	-	7,5	-	5,65
	— to +	5,98	-	7,5	-	5,57
6	m. to —	3,03	2,9	-	-	5,53
	— to +	5,9	5,71		-	5,62
7	m. to —	3,15	3,03	7,4 by mean.	6,57	5,29
	— to +	6,1	5,9			5,44
8	m. to —	3,13	3,00		-	5,34
	— to +	5,72	5,54	-	-	5,79
9	+ to —	6,32	-	6,58	-	5,1
10	+ to —	6,15	-	6,59	-	5,27
11	+ to —	6,07	-	7,1	-	5,39
12	— to +	6,09	-	7,3	-	5,42
13	— to +	6,12	-	7,6	-	5,47
	+ to —	5,97	-	7,7	-	5,63
14	— to +	6,27	-	7,6	-	5,34
	+ to —	6,13	-	7,6	-	5,46
15	— to +	6,34	-	7,7	-	5,3
16	— to +	6,1	-	7,16	-	5,75
17	— to +	5,78	-	7,2	-	5,68
	+ to —	5,64	-	7,3	-	5,85

Figure 10.9 Cavendish's Final Results. *Source*: Cavendish (1798, 520). Reprinted figure from Cavendish. *Philosophical Transactions of the Royal Society* 88: 469–526, 1798.

measurements is 5.48 ± 0.19. The average of all twenty-nine measurements is 5.448 ± 0.22. This discrepancy was first noted by Baily (1843), and later by Poynting and others. They attributed this to an arithmetic error by Cavendish. Baily pointed out that if the third measurement (figure 10.9), published as 4.88, was in fact 5.88 the discrepancy disappears. Baily recalculated the density of the earth using Cavendish's published data for this experiment and found that the value is indeed 4.88.[20]

Using the extreme values Cavendish found in the last twenty-three measurements, he estimated the uncertainty in that value of 0.38. In modern terms we might report the result as 5.48 ± 0.38.[21] "Therefore the density is seen to be determined hereby, to great exactness" (Cavendish 1798, 521).

Cavendish's use of the torsion pendulum was regarded as a significant improvement in the measurements of the earth's density. Previous measurements, as well as some later nineteenth-century experiments, had used a pendulum or a plumb line near a mountain and observed the deflection caused by the gravitational force between the mountain and the plumb bob. Subsequent nineteenth-century experiments also adopted the torsion pendulum and also used pendulums in mines or a common balance. As John Henry Poynting remarked, "[H]e [Cavendish] made the experiment in a manner so admirable that it marks the beginning of a new era in the measurement of small forces" (Poynting 1913, 63).

"G," the Universal Gravitational Constant

Cavendish, as we have seen, measured the density of the Earth, not the universal gravitational constant, "*G*."

From the density of the Earth we may calculate *G* as follows:

$$mg = GmM_{\text{Earth}}/R_{\text{Earth}^2}, G = gR_{\text{Earth}^2}/M_{\text{Earth}} = gR_{\text{Earth}^2}/\left(\left(4/3\right)\pi R_{\text{Earth}^3}\right)\rho_{\text{Earth}}$$

so

$$G = 3g/4\pi R_{\text{Earth}}\rho_{\text{Earth}}$$

where ρ_{Earth} is the density of the Earth.

Using Cavendish's value of 5.48 ± 0.38 for the density of the Earth we find

$$G = \left(6.71 \pm 0.\ 47\right) \times 10^{-11}\,\text{m}^3\text{kg}^{-1}\text{s}^{-2},$$

which is in agreement with the 2014 CODATA value of (6.67408 ± 0. 00031) × 10^{-11} m^3 kg^{-1} s^{-2}.

In 1894, C. V. Boys made a passionate argument for formulating Newton's Law of Gravitation as Force = $G\ (m_1 m_2)/\text{Distance}^2$. He remarked that "*g*," the acceleration due to gravity on the surface of the Earth is

> Eminently of a practical and useful character; it is the delight of the engineer and the practical man; it is not constant, but that he does not mind . . .G, on other hand represents that mighty principle under the influence of which every star, planet and satellite in the universe pursues its allotted course. . . .Owing to the universal character of the constant G, it seems to me to be descending from the sublime to the ridiculous to describe the object of this experiment as finding the mass of the earth, or less accurately, the weight of the earth. (Boys 1894, 330)

Boys's exhortation was somewhat late. As shown by Mackenzie (1900), several scientists had begun to express their results in terms of *G* in the late 1880s and early 1890s. One of these experiments was done by John Henry Poynting (1892), who began his experiment in 1878. Poynting used a common balance and remarked that,

> It might appear useless to add another to the list of determinations, especially when, as Mr. Boys has recently shown, the torsion balance may be used for the experiment with an accuracy quite unattainable by the common balance. But I think that in the case of such a constant as that of gravitation, *where the results have hardly as yet begun to close in on any definite value,*[22] *and where, indeed, we are hardly assured of the constancy itself,*[23] it is important to have as many determinations as possible made by different methods and different instruments, until all the sources of discrepancy are traced and the results agree. (Poynting 1892, 656. Emphasis added)[24]

Poynting took data with his apparatus in two different configurations. "On the completion of Set I the four masses were inverted, and changed over from right to left or left to right, and the initial positions was after this always arranged so that movement of the rider or mass decreased the reading. This was done to lessen errors due to want of symmetry" (Poynting 1892, 600). Poynting averaged his results for the two sets and reported a value of $G = 6.6884 \times 10^{-8}\ \text{cm}^3\ \text{g}^{-1}\ \text{s}^{-2}$ and a density of 5.4934.[25]

Boys followed his own advice and in 1895 published a long paper reporting his experiments measuring *G* using a torsion balance (Boys 1895). He also believed in the variety of evidence. He stated, "I have employed a fair variety of conditions, the lead balls alone being unchanged throughout the series. Three pairs of small masses were made use of" (Boys 1895, 62). His average value for nine different sets of measurements was $6.6630 \times 10^{-8}\ \text{cm}^3\ \text{g}^{-1}\ \text{s}^{-2}$.[26]

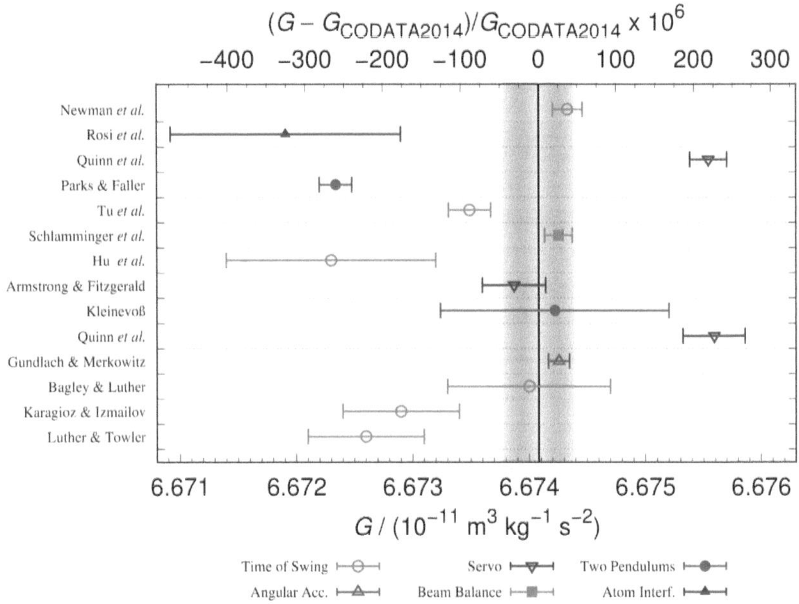

Figure 10.10 Recent Measurements of G. The vertical line shows the recommended value of G and the shading indicates the recommended uncertainty. *Source*: Stephan Schlamminger/NIST.

Is There a Universal Gravitational Constant?

Despite its status as perhaps the most venerable of physical constants,[27] *G*, the universal gravitational constant, is the least well-measured of the important physical constants. As shown in figure 10.10 the most recent measurements of *G* show a wide variation, with many of the measurements reporting values which differ from the accepted value by far more than their stated uncertainties. As Mohr and collaborators remarked in their 2016 review of the recommended values of physical constants, the three then new measurements of *G*, added and obtained by different methods, "have not resolved the considerable disagreements that have existed among the measurements of G for the past 20 years" (Mohr, Newell et al. 2016, 035009-4). The differences are so large that the Task Group assigned to recommend a value for *G* used an expansion factor of 6.3 for the initial uncertainties in the reported values "that reduces the normalized residuals of each datum to less than 2" (Mohr, Newell et al. 2016, 035009-4). The 2014 CODATA value of *G* is $(6.67408 \pm 0.00031) \times 10^{-11}$ m^3 kg^{-1} s^{-2}, with a relative uncertainty of 4.7×10^{-5}. One might contrast this with the uncertainty in the electron *g* factor, the ratio of its magnetic moment to its spin, of $(2.00231930436182 \pm 0.00000000000052)$, with a relative uncertainty of 2.6×10^{-13}.[28]

An attempt to resolve the problem was the 2001 experiment by Quinn and collaborators (2001). They noted that there had been recent measurements of *G* that gave values closer to the CODATA value[29]

> particularly the paper by Gundlach and Merkowitz (2000) that gives a result with the very low uncertainty of 14 ppm [parts per million]. We report here a new determination of G, which has a standard uncertainty of 41 ppm. Our value is unique in that it is based on two results obtained using the same apparatus but with different methods of measurement. Our result does, however, differ from that of Gundlach and Merkowitz by some 200 ppm. (Quinn et al. 2001, 111101-1)

The torsion balance used is shown in figure 10.11.[30] Two different methods were used (1) electrostatic servo control[31] and (2) free deflection (Cavendish method). When the source masses were radially aligned with the test masses the gravitational torque was zero. Rotating the source masses by 18.7° in either direction produced a maximum torque. "In the servo-controlled method, the gravitational torque of the source masses is balanced by an

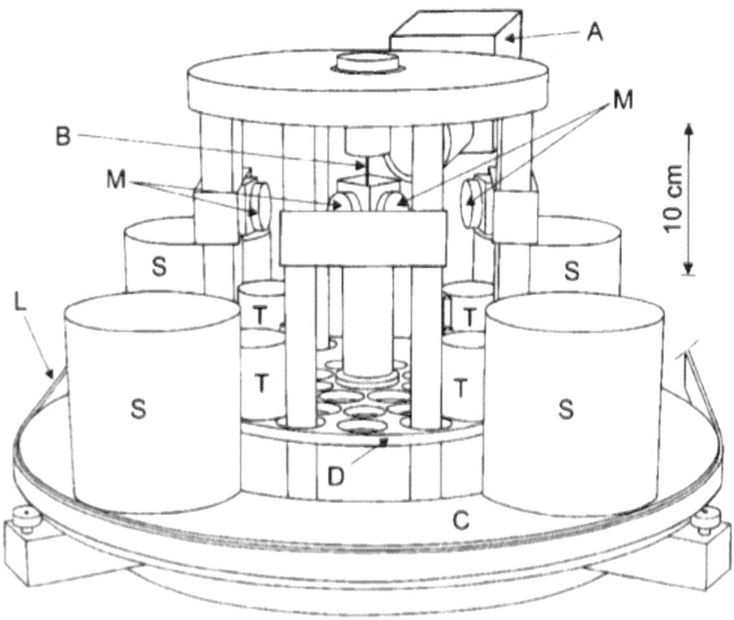

Figure 10.11 The Torsion Balance Used by Quinn and Collaborators. Outline of apparatus: T, test masses; S, source masses; D, torsion balance disk; B, torsion strip. *Source*: Reprinted figure with permission from Quinn, T. J., C. C. Speake, et al., *Physical Review Letters*, 87: 111101-1, 2001. Copyright 2001 by the American Physical Society.

electrostatic torque acting directly on the test masses" (Quinn et al. 2001, 111101-2). In the Cavendish method at equilibrium the applied gravitational torque is balanced by the suspension stiffness, $\tau = \kappa\theta$, where κ is the stiffness of the suspension wire and θ is the angle of equilibrium. Just as Cavendish had done, the stiffness constant was obtained from the period of free oscillation of the test masses and the moment of inertia.

The final result for the servo method was $G = 6.67553 \times 10^{-11}$ m^3 kg^{-1} s^{-2} with a standard uncertainty of 6.0 parts in 10^5. For the Cavendish method $G = 6.67565 \times 10^{-11}$ m^3 kg^{-1} s^{-2} with a standard uncertainty of 6.7 parts in 10^5. The final result was $G = 6.67559(27) \times 10^{-11}$ m^3 kg^{-1} s^{-2} with a standard uncertainty, which included the effects of correlations between the two methods, of 4.1 parts in 10^5.

> In conclusion, the close agreement of the results of our two substantially independent methods is evidence for the absence of many of the systematic errors to which a G measurement is subject. Nevertheless, the two most accurate measurements of G, this one and that of Gundlach and Merkowiz [6.674215 ± 0.000092 $\times 10^{-11}$ m^3 kg^{-1} s^{-2}], differ by more than 4 times their combined standard uncertainty. (figure 10.12) (Quinn et al. 2001, 111101-3)

Gundlach and Merkowitz described the improvements to their apparatus as follows. "We measured Newton's gravitational constant G using a new

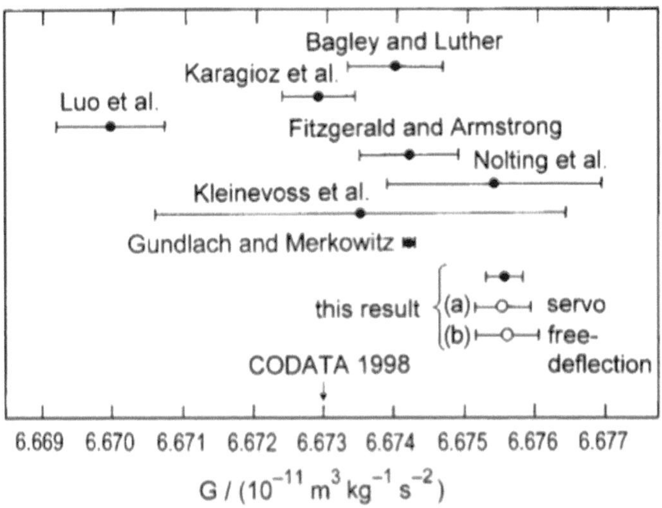

Figure 10.12 Measurements of G Published since 1997, Showing the Values of Gundlach and Merkowitz and That of Quinn et al. (2001). *Source*: Reprinted figure with permission from Quinn, T. J., C. C. Speake, et al., *Physical Review Letters*, 87: 111101-4, 2001. Copyright 2001 by the American Physical Society.

torsion balance method. Our technique greatly reduces several sources of uncertainty compared to previous measurements: (1) It is insensitive to anelastic torsion fiber properties; (2) a flat plate pendulum minimizes the sensitivity due to the pendulum density distribution; (3) continuous attractor rotation reduces background noise" (Gundlach and Merkowitz 2000, 2869). Both results, however, differ considerably from the CODATA 1998 value. The problem remained.[32]

In 2013 Quinn and collaborators reported on their continued efforts to resolve the discordant results. They reported a new value for *G* using the same methods used in their 2001 paper. "The apparatus has been completely rebuilt and extensive tests carried out on the key parameters needed to produce a new value for G" (Quinn, Parks et al. 2013, 101102-1). They further noted that, "The 2010 CODATA evaluation of the fundamental constants shows a spread in the recent values of the Newtonian constant of Gravitation of some 400 ppm, more than ten times the estimated uncertainties of most of the contributing values" (Quinn, Parks et al. 2013, 1011-2-1). Their reported values of *G* were $6.67520(41) \times 10^{-11}$ m^3 kg^{-1} s^{-2} and $6.67566(37) \times 10^{-11}$ m^3 kg^{-1} s^{-2} for the servo and Cavendish methods, respectively. The weighted mean value was $6.67545(18) \times 10^{-11}$ m^3 kg^{-1} s^{-2}. They remarked that their "new value is 21 ppm below our 2001 result which had an uncertainty of 41 ppm but 241 ppm above the CODATA 2010 value" (see figure 10.13) (Quinn, Parks et al. 2013, 101102-4). They further stated, "Noting that in each, the result is based on the average of two largely independent methods, taken together, our two results represent a unique contribution to G determinations" (Quinn, Parks et al. 2013, 101102-5). Nevertheless, the discord has remained to this day.

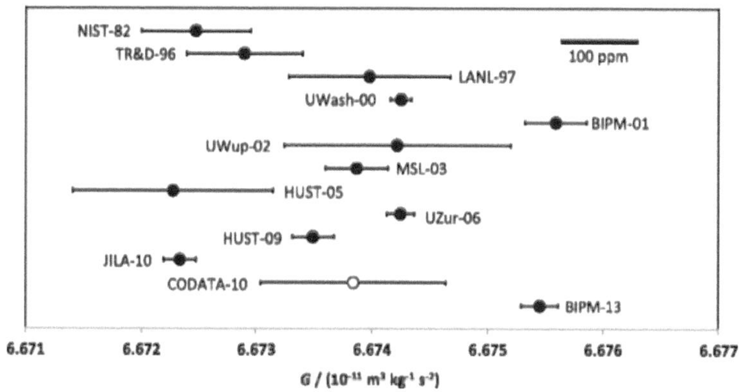

Figure 10.13 Measurements of G, Including the Value of Quinn et al. (2013). *Source*: Quinn et al. (2013). Reprinted figure with permission from Quinn, T., H. Parks, et al., *Physical Review Letters*, 111: 101102-4, 2013. Copyright 2013 by the American Physical Society.

In this episode we observe that considerable technological improvements had been made in the experimental apparatus. Although the claimed precision of the results has improved, the agreement has not. Recall Poynting's comment, "But I think that in the case of such a constant as that of gravitation, *where the results have hardly as yet begun to close in on any definite value, and where, indeed, we are hardly assured of the constancy itself,* it is important to have as many determinations as possible made by different methods and different instruments, until all the sources of discrepancy are traced and the results agree" (Poynting 1892, 656. Emphasis added).

CONCLUSION

In these two episodes we have seen that, in the case of the Universality of Free Fall, improvements in technology have improved the measurements by a factor of 10^{10} and thus allowed more stringent tests of theory. In the case of measurements of G, the universal gravitational constant, measurements from the time of Cavendish to the present have certainly improved in precision, but they do not agree.

There are several interesting questions I have left unanswered. First, why does the torsion pendulum or balance, which is crucial in both episodes, give rise to such different conclusions? I might suggest that in the case of UFF, experimenters are making comparative measurements, whereas in the case of G one is trying to measure a very weak force by itself.[33] In addition, G is not related to any other physical constants. This is not true of other physical constants. For example, the charge of the electron, e, is also involved in the fine structure constant $\alpha = e^2/2\pi hc$, where h is Planck's constant and c is the speed of light. Thus, a measurement of α places a constraint not only on e, but also on h.[34]

The second question concerns the interaction of science and technology. Naively, one might think that science precedes technology. This is clearly not the case. Two recent and very important results, the discovery of the Higgs boson by the CMS and ATLAS collaborations at the Large Hadron Collider and the observation of gravity waves by the LIGO-Virgo collaboration the experimental apparatuses were technological tours de force. The LIGO interferometer is an extraordinarily sensitive complex Michelson interferometer, whose basic design dates back into the late nineteenth century. It uses the interference of light beams which was discovered by Thomas Young in the early nineteenth century. The science came first. The complex detectors of CMS and ATLAS and the powerful computers needed in the analysis of their data required advances in condensed matter physics in their development. In this episode the technology was essential for the science and came first.[35]

The question of which comes first, science or technology, is often difficult to answer, and may very well have different answers in different cases. That remains for future research by a scholar like Joe Pitt, or perhaps even Joe Pitt himself.

NOTES

1. Cooper raises doubt about whether Aristotle actually said this.

2. Galileo did discuss falling bodies several times in his earlier work. For example, in his 1590 treatise, *De Motu*, he states that at the beginning of its fall a wooden body falls more quickly than a lead body. "If the large amount of air in wood made it go quicker, then as long as it is in the air the wood will move ever more quickly. But experience [or experiment] shows the contrary; for, it is true, in the beginning of its motion the wood is carried more rapidly than the lead; but a little later the motion of the lead is so accelerated that it leaves the wood behind; and if they are let go from a high tower, precedes it by a long space; *and I have often made a test of this*" (Galileo, quoted in Cooper 1935, 55). This was written at approximately the time Viviani said that Galileo performed the experiment at the Leaning Tower. The experiments also gave very different results from those described by Galileo in *The Discorsi* and those stated by Viviani.

3. It is also the acceleration due to gravity at the surface of the Earth.

4. After the publication of Einstein's General Theory of Relativity in 1915 this also was referred to as the Weak Equivalence Principle. I will use these terms interchangeably.

5. Eötvös is sometimes credited with the invention of the torsion balance. This is incorrect. As discussed below Henry Cavendish used one in his measurements of the density of the Earth in 1798 and in 1785 Charles Coulomb used one in his investigation of the force between electrical charges.

6. In this discussion I will discuss almost exclusively the work of the Eöt-Wash collaboration. This will, of course, not be a complete history. For details of other work and complete history of the Fifth Force, see Franklin and Fischbach (2016).

7. Based on available evidence, Fischbach et al. (1986) estimated α as approximately 0.01 and λ as approximately 200 m.

8. As seen above it would also have a different distance dependence than the Newtonian force of gravity. I will be concerned here only with the composition dependence.

9. To make matters confusing, the other experimental result reported evidence in favor of the Fifth Force (Thieberger 1987c). For the history of how this discord was resolved, see Franklin and Fischbach (2016).

10. A local mass asymmetry magnified the effect of the Fifth Force.

11. The Eöt-Wash group sometimes referred to their apparatus as a torsion balance and at other times as a torsion pendulum.

12. The results were first presented at the Moriond workshop.

13. Eric Adelberger, one of the senior members of the Washington group, remarked that the point of the *PRL* graph was to show the absence of any Fifth Force

effect, even for a value of α much smaller than that needed for the geophysical or Eötvös data. He noted that, in retrospect, the *PRL* graph did not show this as well as they would have liked, so that in the Moriond paper a more realistic value of α was used (private communication).

14. η is a dimensionless quantity.

15. The Eöt-Wash group continued its whimsy with the naming of their new apparatus.

16. I am grateful to George Smith for helpful discussions on this point.

17. Newton's *Principia* was, of course, published in 1687.

18. Cavendish remarked that the method had been invented by the late Reverend John Michell, who had passed away before he could complete the apparatus and perform any experiments with it. He noted that a similar apparatus had been used by Coulomb, "but Mr. Michell informed me of his intention of making this experiment, and of the method he intended to use, before the publication of any of Mr. Coulomb's experiments" (1798, p. 470).

19. The density of water is 1.00

20. Perhaps Cavendish made an error and actually used 5.88 rather than 4.88.

21. Using a standard deviation calculation for the last twenty-three measurements, Cavendish's result is 5.48 ± 0.19.

22. As we shall see below, this was a prescient remark.

23. During the nineteenth century, the values obtained for the density of the Earth varied from 4.25 to 7.60 (all methods) and from 5.44 to 6.25 (torsion pendulum). The contemporary value is 5.513.

24. For an argument on the value of different experiments, see Franklin and Howson (1984).

25. Modern scientists would report this as 6.6884×10^{-11} m^3 kg^{-1} s^{-2}.

26. Neither Poynting nor Boys reported an experimental uncertainty.

27. Although Newton did not use such a constant in his statement of his Law of Universal Gravitation and though Cavendish measured the density of the Earth and not G, it seems reasonable to consider G as dating from the time of Newton.

28. To be fair, the g factor of the electron is among the best measured physical constants.

29. CODATA values for the physical constants are the accepted values for the physics community.

30. There had been considerable technological improvements in the torsion balance since the end of the nineteenth century. See discussion below.

31. The apparatus is similar to the torsion balance used by the Eöt-Wash collaboration, discussed earlier, except that four source masses have been substituted for the external hill.

32. The experimenters attempted to explain the discrepancy by invoking a failure of Newton's inverse square law of gravity, because the effective distances between the source and test masses were different in the two experiments. They found that the violation required was larger than the limit that had been set by Spero et al. (1980).

33. The electrical force between the electron and the proton in the hydrogen atom is 2×10^{39} larger than the gravitational force between them.

34. The speed of light is now defined to be 299,792,458 m s^{-1}.

35. This is not to say that there was no previous relevant science. The discovery of the Higgs boson was the last remaining piece of the Standard Model, our current theory of elementary particles. The establishment of that theory involved both considerable science and technology.

REFERENCES

Adelberger, E. G., C. W. Stubbs, et al. (1990). "Testing the Equivalence Principle in the Field of the Earth: Particle Physics at Masses below 1μ eV." *Physical Review D* 42: 3267–92.

Baily, F. (1843). "Experiments with the Torsion-Rod for Determining the Mean Density of the Earth." *Memoirs of the Royal Astronomical Society* 14: 1–120 and i.–ccxlviii.

Boys, C. V. (1894). "On the Newtonian Constant of Gravitation." *Nature* 50: 330–34.

Boys, C. V. (1895). "On the Newtonian Constant of Gravitation." *Philosophical Transactions of the Royal Society* 186: 1–72.

Cavendish, H. (1798). "Experiments to Determine the Density of the Earth." *Philosophical Transactions of the Royal Society (London)* 88: 469–526.

Cooper, L. (1935). *Aristotle, Galileo, and the Tower of Pisa*. Port Washington, NY, Kennikat Press.

Eotvos, R. (1890). "Uber Die Anzwiehung Der Erde Auf Verscheidene Substanzen." *Akademie Ertesito* 1: 108–10.

Eotvos, R., D. Pekar, et al. (1922). "Beitrage zum Gesetze der Proportionalitat von Tragheit und Gravitat." *Annelen der Physik (Leipzig)* 68: 11–66.

Fischbach, E., S. H. Aronson, et al. (1986). "Reanalysis of the Eotvos Experiment." *Physical Review Letters* 56: 3–6.

Fischbach, E., D. Sudarsky, et al. (1988). "Long-Range Forces and the Eötvös Experiment." *Annals of Physics* 183: 1–89.

Franklin, A. (1993a). "Discovery, Pursut, and Justification." *Perspectives on Science* 1: 252–84.

Franklin, A. (1993b). "Experimental Questions." *Perspectives on Science* 1: 127–46.

Franklin, A. (1995). "The Resolution of Discordant Results." *Perspectives on Science* 3: 346–420.

Franklin, A. (1997). "Calibration." *Perspectives on Science* 5: 31–80.

Franklin, A. and E. Fischbach. (2016). *The Rise and Fall of the Fifth Force*. Heidelberg, Springer.

Franklin, A. and C. Howson. (1984). "Why Do Scientists Prefer to Vary Their Experiments?" *Studies in History and Philosophy of Science* 15: 51–62.

Galileo. (1974). *Two New Sciences*. Madison, University of Wisconsin Press. Translated by Stillman Drake.

Gundlach, J. and S. M. Merkowitz. (2000). "Measurement of Newton's Constant Using a Torque Balance with Angular Acceleration Feedback." *Physical Review Letters* 85: 2869–72.

Gundlach, J. H., G. L. Smith, et al. (1997). "Short-Range Test of the Equivalence Principle." *Physical Review Letters* 78: 2523–26.

Mackenzie, A. S. (1900). *The Laws of Gravitation; Memoirs by Newton, Bouguer, and Cavendish*. New York, American Book Company.

Mohr, P. J., D. B. Newell, et al. (2016). "CODATA Recommended Values of the Fundamental Physical Constants 2014." *Reviews of Modern Physics* 88: 035009-035001–035009-035073.

Newton, I. (1999). *The Principia, Mathematical Principles of Natural Philosophy*. Berkeley, University of California Press. Translated by I. Bernard Cohen and Anne Whitman.

Poynting, J. H. (1892). "On a Determination of the Mean Density of the Earth and the Gravitation Constant by Means of the Common Balance." *Philosophical Transactions of the Royal Society* 182: 565–656.

Poynting, J. H. (1913). *The Earth: Its Shape, Size, Weight and Spin*. Cambridge, Cambridge University Press.

Quinn, T., H. Parks, et al. (2013). "Improved Determination of *G* Using Two Methods." *Physical Review Letters* 111: 101102-101101–101102-101105.

Quinn, T. J., C. C. Speake, et al. (2001). "A New Determination of G Using Two Methods." *Physical Review Letters* 87: 111101-111101–111101-111104.

Raab, F. J., E. G. Adelberger, et al. (1987). "Search for an Intermediate-Range Interaction: Results of the Eot-Wash I Experiment." In *New and Exotic Phenomena: Seventh Moriond Workshop*. O. Fackler and J. Tran Than Van. Gif sur Yvette, Editions Frontieres.

Schlamminger, S., K. Y. Choi, et al. (2008). "Test of the Equivalence Principle Using a Rotating Torsion Balance." *Physical Review Letters* 100: 041101-041101–041101-041104.

Spero, R., J. K. Hoskins, et al. (1980). "Tests of the Gravitational Inverse-Square Law at Laboratory Distances." *Physical Review Letters* 44: 1645–48.

Stubbs, C. W., E. G. Adelberger, et al. (1987). "Search for an Intermediate-Range Interaction." *Physical Review Letters* 58: 1070–73.

Su, Y., B. R. Heckel, et al. (1994). "New Tests of the Universality of Free Fall." *Physical Review D* 50: 3614–36.

Thieberger, P. (1987). "Search for a Substance-Dependent Force with a New Differential Accelerometer." *Physical Review Letters* 58: 1066–69.

Will, C. (1984). *Was Einstein Right?* New York, Basic Books.

Chapter 11

Joe Pitt, the Philosophical Imagination, and the Practice of Pedagogy

James H. Collier

In 1929, Alfred North Whitehead asserted that imagination is central to the university. Today such an assertion can seem almost quaint, superseded in a university dominated by technological solutionism. Yet, by integrating a philosophy of technological change with an inventive pedagogy, professor Pitt revitalizes Whitehead's vision:

> [T]he proper function of a university is the imaginative acquisition of knowledge. . . . A university is imaginative or it is nothing—at least nothing useful. . . . Imagination is a contagious disease. It cannot be measured by the yard, or weighed by the pound, and then delivered to the students by members of the faculty. It can only be communicated by a faculty whose members themselves wear their learning with imagination. . . . More than two thousand years ago the ancients symbolized learning by a torch passing from hand to hand down the generations. That lighted torch is the imagination of which I speak. (Alfred North Whitehead, *The Aims of Education,* 1929, 96–97)

From his own undergraduate experiences to his scholarship in the philosophy of science and technology, and in his many years as a teacher, professor Pitt wears his learning with imagination.

In a March 25, 2011, interview, Laureano Ralón asked professor Pitt: "How did you decide to become a university professor? Was it a conscious choice?" (p. 1). As part of a longer answer, professor Pitt relays the following story:

> [I]n 1962 . . . when I took my first philosophy course, Frank MacDonald addressed the class during the week in which people were supposed to select their majors.[1] . . . Frank said: "if you want to spend your life trying to solve

> problems that have been asked for 3000 years, and wrestle with people that are much smarter than you and to no avail, then by all means, major in philosophy. But you are doomed to a life of frustration." And he turned around and walked out of the room. (p. 1)

Professor Pitt admits that professor MacDonald set the hook then and there. Having no idea what he would do professionally as a philosophy major, professor Pitt sought a way to integrate his new interest with legal practice. In the same answer, professor Pitt adds: "I was interested in how to use the Law to affect social change, and they [the philosophy faculty at the University of Western Ontario] were working on scientific change, I figured that I might be able to take what I would learn from them and apply it to social change" (p. 1).

How we understand and achieve change—social change, scientific change, conceptual change, Heraclitian change, technological change, philosophical change—recurs as a central concern throughout professor Pitt's scholarship. In the interview above we also see how personal, perhaps even incredulous, change comes about when cast as a challenge by a charismatic teacher. The change wrought by a skilled teacher's imagination results, as Whitehead suggests, not only in learning and knowing but also in continuity and belonging. The philosophical imagination comes alive in a dialectical moment: as we regard tradition, we enact useful change. The philosopher's mind remains vitally a student's mind equipped with the ability to put abstraction into practice. Through professor Pitt program building and pedagogy, he moves skillfully from abstraction to practice. He retains a student's openness to ideas while reflecting on their origin and growth. The ability to occupy different perspectives allows professor Pitt to act with great patience and a generosity of spirit. I know of his generosity. He gives me, and many others, the opportunity and freedom to work through the demands and vicissitudes of a life in philosophy and in the academy. He waits encouragingly as his students stubbornly push back, fall, and get up to push back again. In an exchange (Pitt 2011; Collier 2012) that helped launch an online venture, professor Pitt acknowledged my rather blunt defense of the field of Science and Technology Studies. In similar encounters, professor Pitt discerns arguments with increasingly precise and well-placed examples and questions. In so doing, he wears his learning with imagination.

I have known professor Pitt for roughly forty years. I took several classes with him as both an undergraduate and a graduate student. He chaired my thesis and my dissertation committees. I have attended many of professor Pitt's public lectures, we have participated in academic conference panels together, and served on doctoral committees together. As a faculty member in 2008, I had the unique pleasure of co-teaching a graduate course with him on the "Philosophy of Science and Technology Studies." We challenge and help each other now as we have challenged and helped each other in the past in the classroom, in print, and over meals and drinks. He continues to be my mentor regarding matters large and small, professional and personal. We are colleagues. We are steadfast friends. In this constant relationship, we have witnessed great changes in each other.

In what follows, I sketch out scenes of encounter. These scenes follow a loose dual chronology of my personal history with professor Pitt and of his reactions to what happens in the philosophy of science and the philosophy of technology in the post-Kuhnian era. (1962, when professor Pitt took his first philosophy course, marked the publication of the first edition of the *Structure of Scientific Revolutions.*) I chose these scenes because they demonstrate how professor Pitt embodies the imagination, as defined by Whitehead. Professor Pitt uses his imagination in the service of change, the guiding meta-philosophical principle that binds his scholarship and teaching.

I first met professor Pitt in 1981 after arriving on Virginia Tech's campus in fall 1979. I knew little about academic philosophy on entering the university. Philosophy seemed a particularly exotic, statement-making pursuit (at least to my family and friends) at a land-grant university. I recall a rebellious sense that I alone, among my professionally focused friends, embodied the humanist ethos and the educational goals of a comprehensive research university. I proclaimed my annoyance with continually having to explain what philosophy was (as if I knew), and deal with common question—so, what's *your* philosophy?—yet, at the same time, I enjoyed the initial interest and assumptions regarding philosophy majors. As a philosophy major I was assumed to be serious, thoughtful, deep, reflective, and completely deluded about the how the real world operated. I reveled in others' confusion about philosophy and felt, intimating as much, that I harbored a valuable secret.

On the day during summer orientation when I met with my adviser to select fall classes I insisted, over his strong objection, that I take "Plato's Republic." Being told that first-year students simply did not take such a course added the timeless attraction of forbidden fruit. I recall, on entering the classroom, the smell of the aging building and the profound emotional mixture of fear and excitement. Like professor Pitt's encounter with professor MacDonald, my meeting professor Nicholas Smith left an impression I would neither shake nor fully fathom.[2] I sat engrossed, acutely aware of the bittersweet sense of existing simultaneously outside, and joining in, a great ritual. Performing the ritual, the advanced students posed questions as confident and improbably lengthy provocations. Answers to professor Smith's layered replies came swiftly presented, again, with unnerving confidence. Beyond my bewilderment, I remained drawn to what I took as the continuous promise of philosophy—to come to learn well beneath the surface and then envision how to enter, and perhaps contribute to, a dialogue preceding you by two millennia. I drank in an enchantment of philosophy and, on this feeling alone, imagined what could be more. Professor Smith pointed to the landscape. Professor Pitt would act as the guide.

Professor Pitt cut an imposing intellectual figure to a wide-eyed undergraduate. His assuredness, knowledge, and uncanny ability to connect and refine thinking and argument through Socratic questioning made me, and my classmates, want to learn quickly and think concisely. I witnessed professor Pitt's enthusiasm and joy in presenting and analyzing arguments. I began internalizing as a genuine principle, well beyond the shopworn expression, that *ideas matter*. I was not exactly sure how ideas mattered in the world, but I became convinced that ideas mattered to me. Ideas mattered to me because, initially and in part, they mattered to professor Pitt. He embodied the authority found in diverse, sometimes converging, sets of beliefs. Smart people like professor Pitt dedicated their lives to certain beliefs while often vehemently rejecting other beliefs. Still, I struggled as a student not only with grasping basic philosophical principles, but also with resolving those principles with the nature of the influence of the people holding and teaching them.

My struggle to come to grips with the relation between pedagogical influence and philosophical proficiency, even knowledge, has moved me in long-lasting and disparate ways. As Robert Woodhouse explains regarding Whitehead's ideas on education:

> [I]maginative teaching and learning appeals not only to the intellect but to the emotions, balancing the demands of the former with the energy required for realizing the purposes required to transform reality. In enriching our vision of what *could* be and sustaining the zest capable of enacting that vision, faculty and students need courage to challenge the dominant norms of education and society. (2012, 3)

The balance of intellect and emotion that helps supply the courage necessary to "challenge the dominant norms" and transform reality does not come easily (if at all) or all at once. Professor Pitt likes to relate a story of a moment when, agitated by a particularly spirited exchange, I walked out of class. I do not remember the specific argument or ideas being discussed, but I remember how I felt. I felt frustrated, overwhelmed, and fearful. I could not properly articulate myself in that moment, when my intellect could not keep pace with my emotions, but what was happening was that I was learning. The frustration I experienced was a matter of pedagogical design. As a polished practitioner of the Socratic method, professor Pitt would work out a student's frustration tolerance with complex ideas and arguments. The desire to meet professor Pitt's standard, and the fear in failing to do so, led me to walk out. Of course I returned, determined, for the next class. Yet, from that momentary flash of anger arose great trust. I trusted professor Pitt's intuitions about how to challenge me, and in his inclination to interpret my actions as more than being bratty. He understood me as a student and, so, he tested my potential for handling the difficulties of philosophy. He allowed me to push back. He encouraged it.

As I witnessed over the years, professor Pitt's charisma extended from his desire to meet people where they are. His imaginative relational approach to pedagogy would, among other things, foster the success of his many students, help grow a department, and lead to the creation of a new program.

William Clark's remarkably thorough study *Academic Charisma and the Origins of the Research University* (2006) proves helpful in addressing the common and complicated web of influence among professor, student, and institution as embedded and exercised in academic ritual. The theory of academic charisma Clark employs comes from Max Weber. The original charismatic figure was

> [T]he sorcerer, then later the priest and especially the prophet. . . . Regarding academia, part of academic charisma sprang from this topos—the teacher as spiritual or cultic leader. On the sphere of politics and economics, the original charismatic figure was the warrior, then later the general or king. Part of academic charisma spring from this topos—the martial, agonistic, polemical cast of academic knowledge as it developed in medieval Europe. (Clark 2006, 14–15)

Clark argues that as society and the research university adapted to help achieve the modern bureaucratic state, professorial charisma also changed. The research university transformed the charismatic authority found (and undesired) in highly influential individuals (Peter Abelard serves as an example) into the distributed material rationality found in academic rituals and routines—the ritual of bestowing titles and honors, the routine of running laboratories, of writing and publishing, and of teaching and installing curricula. Even now, the transformed charismatic figure persists and exercises great influence—an influence the university promulgates and exploits.

Many features of the traditional university, including the lecture and the disputation (Clark also identifies them as the sermon and the joust), survive in altered forms today. The traditional disputation, with its theater of oral debate, grew staid and indulgent:

> In 1775, J[ohann] Chladenius remarked that some now said disputation provided no good means to investigate the truth, and that disputation often resolved

> nothing. Chladenius defended oral disputation over polemical writing. In oral disputation the polemics ceased with the actual event, after which the interlocutors were (supposedly) friends again. He made clear that oral disputation served as theater, a play, where the actors only appeared playing roles. In polemical writing, however, one was not playing the same game, since personal reputation stood more at stake. (Clark 2006, 88–89)

Customary disputation has faded into the mists of academic history. However, the features of the joust remained embedded, though transfigured, in the academy and, particularly, in the contemporary adoption of the Socratic method in philosophy.[3] Like many beginning students, I viewed the more or less (too frequently) skilled exercise of "teaching with questions" as a kind of personal grilling absent a tangible outcome. Professor Pitt's rather more artful approach showed what was possible if a teacher regards the academic infrastructure in which these pedagogical tools developed.

I refer to Clark's account as a way to point out the change that surrounded me as an undergraduate student and to think about the pedagogical tools used by professor Pitt. As a student, I adopted particularly to the attitudes and assumptions conveyed by Socratic disputation. Schneider seems troubled in contending that Socratic method

> [W]as re-created and reimagined by different groups of educators who were less concerned with establishing a consistent and specific meaning for the method than they were with using it to advance their own distinct agendas. Thus, . . . the Socratic method is in reality a vaguely defined and relatively modern pedagogical concept—a fact that should give pause to educators presuming to employ it. (Schneider 2013, 613)

Schneider's worries stem from educators mouthing the term "Socratic method" as a way to license all manner of ill-defined instruction by means of questioning. Schneider proclaims that a better way would entail instructors interrogating what they did not know about Socratic method in order to gain a more defined conception about students' "intellectual, moral, and emotional growth" (Schneider 2013, 634). Both Schneider and Clark point out the importance of tracing and questioning the adoption of academic conventions and the material transmission of their values. In a complementary key, professor Pitt shows how in shifting our ground from "abstract philosophical justification . . . to a pragmatic condition of success" (Pitt 2000, 40) we arrive at a grounded analysis and context-based judgment of the conceptual and material tools that we use as scholars and teachers. This process of material expression and transmission lies at the heart of professor Pitt's philosophical and pedagogical imagination.

As both a student and a professional academic, professor Pitt struck me as a charismatic intellectual leader, a person that I wanted to emulate. His style of leadership stemmed from his abilities as a dynamic lecturer. His charisma derived, in part, from his enthusiasm:

> [T]he best teachers are the one who are not only enthusiastic about their subject matter, but manage to convey their enthusiasm to their students. You can present a crystal clear lecture that conveys an incredible amount of information, but if you do it in a monotone way indicating "this is exciting, that is cool" the students are not going to learn. So that's it; enthusiasm about their subject matter. (Ralón 2011, 4)

Let me weave together the aspects of Clark's research with regard to professor Pitt's pedagogy. Clark refers to the charismatic figure of the warrior and the "martial, agonistic, polemical cast of academic knowledge" (p. 15). Clark also refers to the failed practice of oral disputation which gets recast in material form in the nineteenth-century seminar:

> The seminar crucially enhanced the oral culture of academic by compelling students to speak. But the essential charismatic exhibition lay elsewhere. If the original prerequisite for admission to the seminar has been passing an exam, it became complemented and sometimes replaced by the submission of a written paper. . . . The movement toward demanding original writing by no means abolished the medieval technique of formation and evaluation involved in disputation. It rather embraced the disputation as a fundamental practice. (pp. 176–77)

Whitehead speaks of the "imaginative acquisition of knowledge." Professor Pitt imaginatively recasts disputation as a practical problem of how we query taken-for-grant notions about the material world in order to act sensibly and effectively.

I took several classes with professor Pitt. I recall an undergraduate class in which we performed an extended close reading of Nelson Goodman's *Fact, Fiction, and Forecast.* Even given a detailed historical and critical introduction to the issues the text addresses, Goodman's arguments require a focus and patience that students find difficult to sustain. Of course, philosophy students become quite familiar with close reading and the explication of thorny texts. Yet, for many teachers of philosophy the second nature of close reading (as well as scholarly writing) can result in a kind of pedagogical negligence they lack both the ability and the desire to actually teach this fundamental practice to students. Avoiding such negligence requires imagination rooted in empathy.

I remember professor Pitt's insistence that we express the argument in our own words without parroting (with certain allowances) Goodman. Given Goodman's innovative use of time-dependent predicates, professor Pitt's goal struck me as, at best, exceedingly optimistic. Despite our doubts, professor Pitt helped each of us and we learned to encourage, if not help, each other. On the not infrequent occasions when a class member presented a counterargument that missed the mark, professor Pitt would ask other class members to assist by having them restate what they heard and then offer the better formulation. We could disagree, but in the spirit of the principle of charity we were required to offer a stronger version of the argument being made. On professor Pitt's urging, we would persist in trying to turn our argumentative lead into gold. We laughed at ourselves as we tried on each other's views, resulting in a variant of a game of telephone, and sought a place where those views might intersect the text. Professor Pitt promoted this camaraderie as we hiked the long trail up the mountains of "grue" and "bleen."

I would later recognize this lesson as an introduction to conceptual analysis. Throughout the process, professor Pitt's ability to get us to inhabit the text, to express (however woefully) what was happening, and to glimpse the clarity borne by philosophical analysis gave us a sense of "getting it." Like an artisan demonstrating their craft, professor Pitt showed how analytic tools work and what those tools could do in the right hands. At the time, I came away with but a nebulous idea of the problem of induction, and of Goodman's solution. Still, a distinct effect remained. Philosophical inquiry elevated the thought and expression of me and my classmates even if only during the time shared in the classroom. The work of inquiry was neither magical nor unpredictable; rather, we would need to transform what we had taken in and learn to navigate the philosophical landscape.

One might well argue that the 1970s stand as the golden age of the philosophy of science. The publication of the second edition of Thomas Kuhn's *The Structure of Scientific Revolutions* (1970) and of Richard Rorty's *Philosophy and the Mirror of Nature* (1979) provided an unequivocal alpha and omega to this era. While comparisons between these two works warrant circumspection, Kuhn and Rorty's works underscore a common pedagogical challenge faced in the post–golden age era. To wit, if we generally agree that logical empiricist accounts of science and knowledge were mistaken, how might we, going forward, teach the philosophy of science and epistemology to arrive at a more accurate understanding?[4] Following Rorty, a pedagogical way forward might feature multidimensional conversation; that is, students and teachers would convey their ideas in language unrestrained by positivist discourse. Following Kuhn, students and teachers might turn to extra-disciplinary resources, history in particular, to develop a more detailed story about science. With either Rorty or Kuhn as a guiding star, students and teachers confronted the problems that followed in crossing discursive and disciplinary boundaries.

For professor Pitt, as for almost all philosophers of science at the time, Kuhn's influence was keenly felt and variably appreciated. As professor Pitt described:

> "In the *Structure of Scientific Revolutions*, Thomas Kuhn provided a powerful argument philosophers had to acknowledge. . . . An examination of the kinds of social, political, and economic factors Kuhn introduced becomes increasingly important when we try to understand the structure and development of contemporary science" (Pitt 2000, 8).

As I discovered later, professor Pitt and his colleagues throughout the university during the decade after the publication of *Structure's* second edition, busied themselves with confronting Kuhn's challenge. Given the need to examine myriad social and historical factors, how then ought we teach science to our students? How should scholars respond to the increasing awareness of science and technology's impact on society? What new structures should universities put into place to support such scholarship and teaching?

Answers to these questions called for an interdisciplinary approach. The resources of any one discipline seemed insufficient to the task of analyzing science. Moreover, the indulgences of logical empiricism underscored the need for a kind of check and balance system in which portrayals of science might be compared, if not integrated, with one another. The imperative to study science from interdisciplinary perspectives became widely accepted. How to conduct and teach interdisciplinary scholarship, however, remained a contested issue.

Beginning in the mid- to late 1970s, professor Pitt and several colleagues, including the not uncontroversial Henry Bauer, Dean of the College of Arts and Sciences from 1978 to 1986, began to consider a programmatic response to the challenge to the interdisciplinary study of science and technology. Beyond Virginia Tech, history and philosophy of science programs had already gained ground, and the first STS programs sprang up in Europe. I would take a circuitous route to the program. I took my first STS class with professor Pitt in 1986 on the philosophy of science, and would enroll in the program full-time in 1989. I am now a faculty member in the same program.

I took professor Pitt's efforts to start a program that studied science in society as a rare realization of well-conceived humanist impulse. However, turning a good impulse into effective academic practice remained (and remains) no small feat. How, then, to turn an interdisciplinary approach to science into both a curriculum and scholarly practice? In a later conversation with me, professor Pitt shared that the founders of the Center for the Study of Science in Society initially thought a properly trained student would need multiple doctoral degrees; clearly, an impractical approach.[5] A more humane goal involved educating doctoral candidates to the level of possessing a working idea of how research in a field addressed scientific issues so that they might attend a respective conference and participate fully. Yet, curricular issues persisted. How does one teach interdisciplinarity? What does good interdisciplinary scholarship look like? How might students satisfy the high, and highly varied, standards of scholarship found in, at least, history, philosophy, and sociology? Would the sum of the scholarly parts be less than the whole? To answer these questions, professor Pitt and the Center's faculty would call on students to embody interdisciplinary synthesis and change. I was one of those students.

In "Against the Perennial: Small Steps toward a Heraclitian Philosophy of Science," professor Pitt addresses the issue of embodiment with regard to technological artifacts. Professor Pitt argues that if we see material technologies as "embodying" knowledge, then conceiving of knowledge as justified true belief, as we have perennially, is insufficient. The Cartesian legacy that continues to dog how we define and think about knowledge does not offer the conceptual resources we need to pose and answer questions about its shifting material expression.

> To reconceptualize the issue by rejecting the Cartesian dualism and embracing a form of naturalism *changes the philosophical landscape,* as long as we actually reject the old and embrace the new. The new view takes human action and its product as primitive and concentrates on the processes by which goals, desires and values are transformed by those actions and products. (Pitt 2003, 11. Emphasis added)

The metaphor of the changing philosophical landscape occupies a central place as professor Pitt explains in the article's opening: "I first offer a metaphor through which I propose we conceptualize the dynamics of philosophical change. This is the metaphor of landscapes, continents and plate tectonics" (2003, 2). The philosophy of technology serves as the harbinger of this change. By forgoing perennial problems and, so, a fixed route, and by analyzing the complex contexts in which humans make objects, we continually reorient ourselves to our surroundings. Moreover, we have the opportunity to ascend to a place where we can appreciate what lies before us.

In helping establish Center for the Study of Science in Society and, so the Science and Technology Studies (STS) department at Virginia Tech, professor Pitt sought a place to reorient inquiry on science and technology by seeking a second-order transformation. As professor Pitt explains:

> A second-order transformation involves a constructed device. An oil refinery performs a second-order transformation; so does a legal system or a geometry or a telescope. . . . [U]sing the basic notion of an input/output process, we can still distinguish among mechanical, social, and decision-making processes, thereby allowing such institutionalized decision-making processes as bureaucracies and funding agencies to be characterized as technologies. (2000, 13–14)

In describing how to think about technology, professor Pitt advocates for the epistemic priority of technology in relation to science. He also locates constructed devices as second-order transformations. To backtrack a bit, professor Pitt gives examples of first-order transformations as being a set of deliberations, or a decision or decisions. A third component of the process is a formal assessment process which feed back into the decision-making process—a form of correction.

I offer this brief summary of professor Pitt's model of technology as a way to think about the construction of STS at Virginia Tech. The constructed "device" in this instance was an academic center and then a department. The normative ideal would be to take first-order transformations, deliberations about a scientific theory or technological artifact, for example, and take those deliberations as input for scholarly and pedagogical output. In theory that output could also serve as an assessment component. In this way, the STS department might act like the academic iteration of the United States Office of Technology Assessment (1979–1995).[6]

Of course, one of the outputs of a doctoral program are students. Individual students, especially those located in an interdisciplinary academic infrastructure, are a lovely way to muck up a perfectly designed course or program—even if such a program is designed as we remain "alert to the complications that the technological infrastructure creates" (2000, 138).

In the Preface of *Pictures, Images, and Conceptual Change,* professor Pitt outlines the challenges of assessing Wilfrid Sellars's work (primarily in relation to the rationality of scientific change). The study of Sellars's work poses a problem in that he "is a philosopher in the very best sense of the word. . . . He performs the task of analyzing alternative views with both finesse and insight . . . [and] he is a systematic philosopher" (Pitt 1981, vii). Pointing out Sellars's virtues, virtues shared by canonically anointed philosophers, professor Pitt invites questions as to how one should treat "past builders of philosophic systems" (Pitt 1981, vii). One can, for example, hunt for inconsistencies in the system, scrutinize presuppositions, or make evaluations based on the guideposts found in particular schools of philosophic thought (Pitt 1981, vii–viii). Sellars's distinctiveness, on professor Pitt's rendering, frustrates these approaches. In order to fruitfully explore Sellars's work, professor Pitt limits the scope of his inquiry to the possible future of scientific development based on an idea of conceptual frameworks and their relation to the world.

Professor Pitt poses Sellars as "a philosopher in the very best sense of the word." I, admittedly, know nothing of Sellars as a teacher. Moreover, I might be tempted to argue with professor Pitt that one cannot get an adequate sense of "a philosopher in the very best sense of the word" without accounting for them as a teacher. To paraphrase Whitehead, a philosopher is imaginative or they are nothing. In his ability to imagine change in its many particular forms, professor Pitt binds together the elements of his scholarship, teaching, and academic citizenry into a philosophic system that creates a material world in which people inhabit—in which I inhabit. I, one among professor Pitt's many students, will take the torch. We will hand it on.

NOTES

1. A January 8, 2006, obituary of professor MacDonald in the *Daily Press* (Newport News, Virginia) provides a telling insight regarding professor Pitt's own influences and trajectory: "[Frank MacDonald was] one of the most effective and well-loved teachers at William and Mary. Hundreds of students at a time attended his deservedly popular lectures on the History of Philosophy where they savored Frank's eloquent, erudite and witty commentaries on the great books of the western intellectual tradition. [Professor Alan] Fuchs further cites the insightful and delightful contribution that MacDonald provided during meetings of the Faculty of Arts and Sciences, and to his ability to draw upon hundreds of poems that he had memorized and recited at suitable moments. Frank retained this startling ability into the last days of his long life."

2. Professor Smith is currently the James F. Miller professor of humanities, and professor of philosophy, at Lewis and Clark College. His website begins: "My goal in every class I teach is to try to infect my students with the same fascination and passion for philosophy and the classics that I have always felt for them. Teaching, for me, is an attempt to share my love for these subjects." https://college.lclark.edu/live/profiles/77-nicholas-smith.

3. The unceasing tales of Sidney Morgenbesser's apothegms in serving up devastating refutations of arguments lends evidence to the high regard the joust still retains in philosophical culture.

4. I take this point from George Reisch: "Most contemporary philosophers, however much they may appreciate logical empiricism as their profession's founding movement, agree that in the 1950s and '60s logical empiricism was revealed to be a catalog of mistakes, misjudgments, and oversimplifications about science and epistemology" (2005, 1).

5. Only somewhat more sensibly, the working idea seemed to be that a student might meet the criteria of having multiple doctorates through some combination

of examinations and research output (articles, dissertation) absent the equivalent coursework.

6. "The Office of Technology Assessment (OTA) was an office of the United States Congress from 1972 to 1995. OTA's purpose was to provide Congressional members and committees with objective and authoritative analysis of the complex scientific and technical issues of the late 20th century, i.e. technology assessment. It was a leader in practicing and encouraging delivery of public services in innovative and inexpensive ways, including early involvement in the distribution of government documents through electronic publishing. Its model was widely copied around the world." https://en.wikipedia.org/wiki/Office_of_Technology_Assessment.

REFERENCES

Clark, William. 2006. *Academic Charisma and the Origins of the Research University*. Chicago: University of Chicago Press.

Collier, Jim. 2012. "Normativity and Nostalgia: A Reply to Pitt." *Social Epistemology Review and Reply Collective* 1 (2): 26–34.

Daily Press. Newport News, Virginia. January 8, 2006. "Frank Aborn MacDonald." https://www.dailypress.com/news/dp-xpm-20060108-2006-01-08-0601080102-story.html.

Goodman, Nelson. 1955. *Fact, Fiction, and Forecast*. Cambridge, MA: Harvard University Press.

Kuhn, Thomas. 1970. *The Structure of Scientific Revolutions*. Chicago: University of Chicago Press.

Pitt, Joseph C. 1981. *Pictures, Images, and Conceptual Change: An Analysis of Wilfrid Sellars' Philosophy of Science*. Dordrecht: Springer.

Pitt, Joseph C. 2000. *Thinking About Technology*. New York. Seven Bridges Press.

Pitt, Joseph C. 2003. "Against the Perennial: Small Steps toward a Heraclitian Philosophy of Science." *Techné* 7 (2): 1–13.

Pitt, Joseph C. 2011. "Standards in Science and Technology Studies." *Social Epistemology Review and Reply Collective* 1 (1): 25–38.

Ral6n, Laureano. "Interview with Joseph Pitt." *Figure/Ground*, March 25, 2011, 1–7. http://figureground.org/.

Reisch, George A. 2005. *How the Cold War Transformed Philosophy of Science: To the Icy Slopes of Logic*. Cambridge: Cambridge University Press.

Rorty, Richard. 1979. *Philosophy and the Mirror of Nature*. Princeton, NJ: Princeton University Press.

Schneider, Jack. 2013. "Remembrance of Things Past: A History of the Socratic Method in the United States." *Curriculum Inquiry* 43 (5): 613–40.

Whitehead, Alfred North. 1929. *The Aims of Education*. New York: The Free Press.

Woodhouse, Howard Robert. 2012. "The Courage to Teach: Whitehead, Emotion, and the Adventures of Ideas." *Collected Essays on Learning and Teaching* 5: 1–5. doi: https://doi.org/10.22329/celt.v5i0.3353.

Afterword

Joseph C. Pitt

First of all, my heartfelt thanks to everyone who contributed to this volume. When Ashley told me what she and Andrew were up to I was, frankly, embarrassed. They and you have much more important things to do than think about my stuff. That said, I am also highly flattered that you would take the time to write something. Each of these papers has made me sit down and think; I am so appreciative of the stimulation! However, I have decided not to respond to each of these papers, as that would take forever. What I will do is reflect on what I have done over the years, triggered by what you have written.

When I first went to graduate school, I had no idea what would come after. During graduate school, there was no preparation for the world after. Yes, we supposed we would get jobs teaching at a university. But no one taught us how to teach. And while we supposed that we would have to publish, having heard many times the old adage "publish or perish," we really didn't know what that entailed. I was fortunate to have a piece deriving from a course final paper accepted at *Kant Studies*. It was a critique of a paper Nick Rescher had published. But it got me started, and it helped me get to know some of the ropes. What I did know was that I liked to teach—having had the opportunity to teach my own course my last year at Western. I just didn't know if I would be any good at it.

I also gave very little thought to what my students would go on to do. When I came to Virginia Tech, there were no graduate programs I could participate in, so I only had undergraduates to worry about. They were, and are, fun and a few went on to graduate school, but most, as is always the case, just wanted to get a job and get on with their lives. No problem there. But then we developed some graduate programs; first, the Science and Technology Studies MS and PhD programs and then an MA in philosophy. Now, suddenly, the futures of these students became of paramount interest. Had I prepared them to succeed in the classroom and in the world of their professions? Some of the essays here reveal surprising success to me. So, maybe I did do something right!

There was also another side of my professional life I had not considered: that I was to be a professional philosopher. From graduate school I knew that participating in professional conferences was supposed to be important, but I didn't really know why. But I started submitting papers and got on some programs and started making some friends outside the small circle of my department and those I went to graduate school with. Sometimes these friendships started by accident, as when I met Ray Dacey at a philosophy of science meeting in the early 1970s. He had not made a room reservation and he was standing in front of me at the hotel desk and I volunteered my extra bed in my room. What I discovered was that being a member of the profession and interacting with friends from around the world made doing philosophy a lot more interesting and fun. But I still hadn't found my niche—I was simply writing papers for the sake of getting on a program or getting a publication.

One day in the mid-1960s, after dropping my wife off at her office, I found myself following a pick-up truck with a bumper sticker that read, "Guns don't kill, people do. Join the NRA." Something about that bumper sticker struck me as right and something struck me as wrong. I said to myself, surely there must be some philosophers working on this issue and off I went looking for philosophers talking about technology, as I soon found out what the landscape was about. I went to my first meeting of the Society for Philosophy and Technology at an American Philosophical Association conference and came way both angry and confused. These people were all Luddites. Surely there was more to say about technology than "it is bad and ruining our lives." As I continued to attend these meetings, I found a few others who thought that there were more interesting issues to pursue, like the nature of technological knowledge and the relationship between science and technology. Pursuing these issues led me away from analytic philosophy of science. As I found out more and more about the kinds of technologies involved in doing science, I became more disaffected from abstract discussions of the nature of probability. I had found my niche. I don't think this sort of volume would have been possible if I hadn't carved my way into the philosophy of technology and found students interested and eager to push the envelope. Equally encouraging still were my colleagues who elected to engage with me on these issues. As usual, the push to write about these issues came from teaching about them. My undergraduates demanded answers which I didn't have. So, when that happened, I turned to writing, which led to *Thinking about Technology*, which led to further engagement with the broader philosophical community.

What was crucial to my development within philosophy of technology was my participation in the Society for Philosophy and Technology. Founded in the 1970s, the members of the Society were equally instrumental as my students in the evolution of my ideas. To begin with, folks like Carl Mitcham and Paul Durbin were not working in the analytic tradition, so I had to convince them that I had something to say. Durbin was not convinced, but Carl developed into a much valued friend and interlocutor. Likewise Stan Carpenter, who was the Society's secretary for many years. Well, I must have done something of value as I was elected to the Board of Directors of the Society and eventually became its president. After that I went on to edit the Society's journal, with the assistance of Ashley Shew. One of the nicest things about the Society for Philosophy and Technology is the atmosphere at the meetings. This was what I imagined professional meetings to be like: a gathering of a group of friends who had a common topic they wanted to discuss with each other. It is friendly, welcoming, vigorous, international, and something I look forward to.

The one idea that keeps making its way into my thinking and writing is that of a feedback loop. It came out of my efforts to develop a pragmatic epistemology—one that centered knowledge in action. But actions often fail to produce the results we want, and so we head back to the drawing board. This approach to epistemology grew out of my dissatisfaction teaching a standard epistemology course to undergraduates. They could not get enthusiastic about justified true belief, and neither could I. So, we started in on a journey to find an epistemology that connected with my undergraduates and eventually my graduate students. It always comes back to the students and they then teach me. That is the best kind of feedback loop.

Index

Pages in *italic* indicate figures.

About the Contributors

Brandiff R. Caron is assistant professor in the Centre for Engineering in Society at Concordia University in Montreal, Quebec. He received his PhD in science and technology studies in 2012 from Virginia Tech where Joe Pitt served as his advisor. Brandiff regularly teaches courses such as the "Social and Ethical Implications of Information and Communication Technologies" and "Impact of Technology on Society" to undergraduate engineering and computer science students. His research has shifted from technology assessment and public policy to technology assessment and engineering pedagogy. His current work focuses on how to best incorporate constructivist technology assessment mechanisms into open-ended engineering design courses.

James H. Collier is associate professor of science, technology and society at Virginia Tech. He is the series founder and editor of "Collective Studies in Knowledge and Society" published by Rowman & Littlefield International, the founding editor of the *Social Epistemology Review and Reply Collective* (https://social-epistemology.com/), and the executive editor of *Social Epistemology: A Journal of Knowledge, Culture and Policy.* Recently, he published a lengthy essay review on Steve Fuller's influence and recent contributions to social epistemology—"Social Epistemology for the One and the Many" (2018) for the *Social Epistemology Review and Reply Collective*. His most recent edited book is *The Future of Social Epistemology: A Collective Vision* (2015)—a volume written collaboratively with the *Social Epistemology Review and Reply Collective.*

Anne C. Fitzpatrick is currently senior cyber and technology risk analyst for the US Department of Defense. She was previously the associate director for Strategic Initiatives at the Department of Defense High Performance Computing Modernization Program. Before this she served on a joint duty assignment as the deputy assistant project director for Future Computing at the National Security Agency, where she led exploration on trends in the next generations of supercomputing architecture and concepts; during that tenure, she was on loan from the U.S. Department of Energy, where she was a senior science and technology intelligence analyst at the DOE Office of Intelligence and Counterintelligence. She has also worked as a technical staff member at Los Alamos National Laboratory. Fitzpatrick gained a PhD in Science and Technology Studies from Virginia Tech in 1998. She also holds a BA and MA from Virginia Tech, and a 1993 Russian Language Proficiency degree from the University of St. Petersburg, Russia.

Allan Franklin is professor of physics, emeritus, at the University of Colorado. He began his career as an experimental high-energy physicist and later changed his research area to history and philosophy of science, particularly on the roles of experiment. He has twice been chair of the Forum on the History of Physics of the American Physical Society and served two terms on the Executive Council of the Philosophy of Science Association. In 2016, Franklin received the Abraham Pais Prize for History of Physics from the American Physical Society. He is the author of eleven books, including most recently *Shifting Standards: Experiments in Particle Physics in the Twentieth Century* and *What Makes a Good Experiment? Reasons and Roles in Science*.

Andrew Wells Garnar has been known to teach at the University of Arizona. He currently works with the tools of American pragmatism to understand the social implications of technology. He also dabbles in the philosophy of science. He earned his MS and PhD from Virginia Tech, both under the supervision of Joe Pitt. He has not made Joe regret this. Yet.

Ivan Guajardo is associate professor of philosophy in the College of Liberal Arts and Social Sciences at Virginia Western Community College, where he also serves as the faculty sponsor for the Philosophy Club and the Human Rights Club. Ivan works in the areas of metaphysics, philosophical anthropology, applied philosophy, imagination, and philosophy of technology.

Peter Kroes has a degree in physical engineering from the University of Technology Eindhoven and a PhD in the Philosophy of Physics from University of Nijmegen (The Netherlands). He is emeritus professor in philosophy of technology at the University of Technology Delft. His main areas of interest are philosophy of technology and philosophy of science. Recent book publications: *Artefact Kinds: Ontology and the Human-Made*

World, eds. Maarten Franssen, Peter Kroes, Thomas A. C. Reydon, and Pieter Vermaas, Synthese Library 365 (2013) and *Technical Artefacts: Creations of Mind and Matter* (2012).

Ronald Laymon was a professor of philosophy specializing in the philosophy of science at the Ohio State University from 1970 to 1995. He took advantage of an early retirement option and completed a law degree at the University of Chicago School of Law in 1997. He then went on to practice large-scale commercial litigation and retired from the full-time practice of law in 2007. While practicing, he had the good fortune to serve as second chair on a case before the U.S. Supreme Court. After retirement from the full-time practice of law, Ron provided pro bono legal assistance to various nonprofit organizations. He currently does consulting for a biotech, intellectual property firm that facilitates the open source creation of therapeutic technologies for use by the global scientific community.

Joseph C. Pitt received his AB in Philosophy from the College of William and Mary in 1966 and his MA (Philosophy, 1970) and PhD (Philosophy, 1972) from the University of Western Ontario in London, Ontario, Canada. He arrived at Virginia Tech in 1971 and, except for some visits to the University of Pittsburgh Center for the Philosophy of Science, he has pursued his academic career there. In 1978, he developed and then for ten years directed the Humanities, Science and Technology Program in the Center for Programs in the Humanities. In 1980, he became the founding director of the Center for the Study of Science in Society. In 1990, he assumed the headship of the Department of Philosophy, stepping down in 1997, returning to the job in 2001 to finally be set free in 2007 and then again from 2012 to 2014. Dr. Pitt has received several teaching awards, including the Alumni Teaching Award and he is a member of Virginia Tech's Academy of Teaching Excellence, which he chaired in 1980–1981. He has authored four books, edited or co-edited thirteen, and published over fifty articles and numerous book reviews. A past president of the Society for Philosophy and Technology, he was also the founding editor of the journal *Perspectives on Science: Historical, Philosophical, Social*, published by the MIT Press and was also editor in chief of *Techné: Research in Philosophy and Technology*. In their spare time, Joe and his wife Donna raise Irish wolfhounds and three-day event horses on their farm, Calyddon, in Newport, Virginia. His most recent book is *Heraclitus Redux: Technological Infrastructures and Scientific Change*.

Nicholas Rescher is a distinguished university professor of philosophy at the University of Pittsburgh. In a productive research career extending over six decades he has well over one hundred books to his credit. Fourteen books about Rescher's philosophy have been published in five languages. He has served as a president of the American Philosophical Association, the American Catholic Philosophy Association, the American G. W. Leibniz Society, the C. S. Peirce Society, and the American Metaphysical Society, as well as the secretary general of the International Union of History and Philosophy of Sciences. Rescher has been elected to membership in the American Academy of Arts and Sciences, the Academia Europea, the Royal Society of Canada, and the Royal Asiatic Society of Great Britain. He has been awarded the Alexander von Humboldt Prize for Humanistic Scholarship in 1984, the Belgian Prix Mercier in 2005, the Aquinas Medal of the American Catholic Philosophical Association in 2007, the Founder's Medal of the Metaphysical Society of America in 2016, and the Helmholtz Medal of the Germany Academy of Sciences (Berlin/Brandenburg) in 2016. In 2011, he was awarded the premier cross of the Order of Merit (*Bundesverdienstkreuz Erster Klasse*) of the Federal Republic of Germany, and honorary degrees have been awarded to him by eight universities on three continents. In 2010, the University of Pittsburgh honored him with the inauguration of a biennial Rescher Medal for distinguished lifetime contributions to systematic philosophy and in 2018 the American Philosophical Association launched a Rescher Prize with a similar objective.

Ashley Shew, MA, MS, PhD, serves as an associate professor at Virginia Tech in the Department of Science, Technology, and Society. Trained by Joseph C. Pitt in philosophy of technology, Shew takes this expertise into her work on emerging technologies, disability studies, and animal studies. She is the co-editor, with Pitt, of *Spaces for the Future: A Companion to Philosophy of Technology* (2017) and sole author of *Animal Constructions and Technological Knowledge* (2017).

Thomas W. Staley is collegiate associate professor in the Department of Materials Science and Engineering, and an affiliated faculty member with the Department of Science, Technology and Society at Virginia Tech. When not entirely occupied with his role teaching aspiring engineers, his research interests include the relationship of the human and social sciences to formal philosophy in the modern era; the history of the human senses and sensory technologies; and the ethical and cultural dimensions of technology and engineering. He is also keenly interested in music and cooking, both as producer and consumer.

www.ingramcontent.com/pod-product-compliance
Ingram Content Group UK Ltd.
Pitfield, Milton Keynes, MK11 3LW, UK
UKHW041842031026
15587UKWH00002B/955

9 781498 597623